42nd Annual Meeting of the Plant Growth Regulation Society of America 2015

Kona, Hawaii, USA
19-23 July 2015

ISBN: 978-1-5108-2885-8

Printed from e-media with permission by:

Curran Associates, Inc.
57 Morehouse Lane
Red Hook, NY 12571

Some format issues inherent in the e-media version may also appear in this print version.

Printed by Curran Associates, Inc. (2016)

Plant Growth and Regulation Society of America
1018 Duke Street
Alexandria, VA 22314
USA

Phone: 703-836-4606 x116
Fax: 703-836-4607

cslone@ashs.org

Additional copies of this publication are available from:

Curran Associates, Inc.
57 Morehouse Lane
Red Hook, NY 12571 USA
Phone: 845-758-0400
Fax: 845-758-2633
Email: curran@proceedings.com
Web: www.proceedings.com

MEETING SUMMARY

42nd Annual Meeting of the Plant Growth Regulation Society of America

The 42nd annual meeting of the Plant Growth Regulation Society of America was held in beautiful Kona, Hawaii. PGRSA meeting participants had the opportunity to see coffee, cocoa, vanilla, and an incredible array of other tropical crops in a wonderfully relaxing setting. The 2015 meeting of PGRSA was a joint meeting with The Plant Growth Substance Special Committee in The Chinese Society of Plant Biology.

The 2015 meeting included an excellent lineup of invited and contributed presentations across a wide range of topics related to plant growth regulation. The meeting included Keynote presentations by Bill Davies, Lancaster University; Daoxin Xie, Tsinghua University; Leon Kochian, USDA-ARS; Karl-Friedrich Bangerth, University of Hohenheim; Andrew Krajewski, International Citrus Technologies Pty Ltd.; Jozsef Racsko, Valent BioSciences Corporation; Hong-Wei Xu, National Key Laboratory of Plant Molecular Genetics; and Marcus Meadows-Smith, Bioconsortia Inc..

Eight students participated in the meeting and four received travel awards. The best Student Oral Presentation Award was presented to Estefania Vincenti, University of California-Davis. The Best Student Poster Presentation award was presented to Mohamad Golam Mostafa, Bangabandhu Shiekh Rahman Agricultural University. The PGRSA Historical Perspectives Award was presented to Karl-Friedrich Bangerth, University of Hohenheim and the PGRSA Young Scientist Award was presented to Jozsef Racsko, Valent BioSciences Corporation.

The PGRSA program committee members including Holly Little, Tim Spann, Carl Sams, Michael Rethwisch, Chris Gunter, Jozsef Racsko, Gurreet Brar, Sonja Maki and Casey Barickman are to be commended for their work in developing this year's program.

On behalf of the PGRSA Steering Committee, we invite all who have an interest in plant growth regulation to attend the 43rd Annual PGRSA meeting in Raleigh, NC.

Aloha!

Carl E. Sams

PGRSA President

HONORARY LIFETIME MEMBERS

Anson Cook

Aldo J. Crovetti

Eric A. Curry

Horace G. Cutler

Richard Dunand

C. Dave Fritz

Masayuki Katsumi (JSCRP)

Wayne Mackay

Wilhelm Rademacher

Edward F. Sullivan

Acknowledgements

The Proceedings has been published yearly since the fourth annual meeting in 1977 at no charge to the author(s). Papers in this edition were received electronically and, although we have made minor changes in formatting, every effort has been made to preserve the original submissions, the content of which remains the sole responsibility of the respective author(s).

The Proceedings were compiled by:
Dr. Holly Little, 2015 to 2016 President

Acadian Seaplants Ltd.
2443 Fair Oaks Blvd., #67
Sacramento, CA 95825
530-219-5415 (phone)
hlittle@acadian.ca

The Proceedings are available for download free of charge at www.PGRSA.org.

Plant Growth Regulation Society of America
Ms. Cindy Slone, Executive Secretary
1018 Duke Street
Alexandria, VA 22314
Tel: 703-836-4606
FAX: 703-836-4607
cslone@ashs.org

A SPECIAL THANK YOU TO THE FOLLOWING COMPANIES FOR FINANCIAL SUPPORT OF THE MEETING

PRESENTING SPONSOR

VALENT BIOSCIENCES CORPORATION

PLATINUM SPONSOR

StollerUSA
Unleashing The Power Of Plants

SILVER SPONSORS

ohp Partners with solutions

Acadian Seaplants

fine

BRONZE SPONSORS

GROW MORE
A Symbol of Quality

Ocean Organics

2014 – 2015 PGRSA Steering Committee

Dr. Carl Sams (President)

The University of Tennessee
Department of Plant Sciences
Rm 252 Ellington Plant Sciences Bldg.
2431 Joe Johnson Drive
Knoxville, TN 37996-4561
865.974.8818 (phone)
865.974.1947 (fax)
carlsams@utk.edu

Dr. Tim Spann (Bus. Manager, Past President)

California Avocado Commission
12 Mauchly, Suite L
Irvine, CA 9261
949.754.0732 (phone)
949.208.3512 (fax)
tspann@avocado.org

Dr. Holly Little (First Vice President)

Acadian Seaplants Ltd.
2443 Fair Oaks Blvd., #67
Sacramento, CA 95825
530-219-5415 (phone)
hlittle@acadian.ca

Mr. Michael Rethwisch (Second Vice President)

University of Nebraska—Lincoln
451 N. 5th Street
David City, NE 68632
402.367.7410 (office)
mrethwisch2@unl.edu

Dr. Chris Gunter (Secretary)
Horticultural Science
North Carolina State University
230 Kilgore Hall, Box 7609
Raleigh, NC 27619-7609
919-513-2807 (phone)
919-515-2505 (fax)
chris_gunter@ncsu.edu

Dr. Casey Barickman (MAL-1)

North Mississippi Research and Extension Center
Mississippi State University
PO Box 1690
Verona, MS 38879
Office 662-566-8001
tbarickm@ext.msstate.edu

Dr. Jozsef Racsko (MAL-2)

Valent Biosciences Corp.
870 Technology Way
Libertyville, IL 60048
Office 847-968-4720
Cell 224-456-1366
jracs@valent.com

Dr. Gurreet Brar (MAL-3)

Farm Advisor (Nut Crops)
University of California Cooperative Extension
Serving Fresno & Madera Counties, California
Office 559-241-7526 (Fresno)
Office 559-675-7879 Ext. 7209 (Madera)
gurbrar@ucanr.edu

Dr. Langtao Xiao (Adjunct Member)
Hunan Agricultural University
Key Lab of Phytohormones
Changsha, Hunan
CHINA 410128
86-731-84635261 (phone)
86-731-4635261 (fax)
langtaoxiao@163.com

Dr. Sonja L. Maki (Executive Director)
Biology Department
Carleton College
1 North College Street
Northfield, MN 55057
507.222.4544 (phone)
507-222-5757 (fax)
smaki@carleton.edu

Ms. Dawn Mancini (Executive Secretary)

PGRSA/American Society for Horticultural Science
1018 Duke Street
Alexandria, VA 22314
703-836-4606 x113 (phone)
703-836-2024 (fax)
dmancini@ashs.org

Table of Contents

ENHANCING GENETIC GAINS IN YIELD THROUGH EXPLOITATION OF NOVEL HORMONE BIOLOGY UNDER HEAT AND DROUGHT STRESS

Bill Davies

Lancaster University, Lancaster, UK

We address genotype x environment x management interactions in the field to understand how any trait can have positive, negative or neutral effects, depending on the nature of the stress challenge. High-throughput PGR-trait phenotyping allows us to understand and exploit the involvement of Plant Growth Regulators in the GxExM interaction. Work on management exploitation of hormone relations is described. More recent work focusses on defining genetic variation in the chemical regulation of both plant productivity and susceptibility to stress. We consider regulation of development of both vegetative and reproductive plant parts under heat and drought and the exploitation of novel understanding of the involvement PGRs in this regulation.

To advance this aim, elite genetic resource panels, a mixture of primitive species, landraces, advanced lines and primary synthetics are phenotyped under heat and drought stress in target environments. Quantitative comparative approaches are used to define genetic potential (new stress indices incorporating both estimates of productivity and susceptibility). High-throughput hormone quantification techniques are applied at an organ, plant and plot scale.

Our ultimate aim is to incorporate PGR-defined traits (growth, development, yield and physiology) into crop improvement programs through pre-breeding. PGR-defined progeny will be tested in target environments and promising material specific to stress environments (e.g. moderate/severe heat, and moderate/severe drought) will be identified and assimilated into other pre-breeding activities, targeting international nurseries and national wheat programs.

REGULATION OF PLANT AUXIN RESPONSE BY IBR5, A DUAL SPECIFICITY PHOSPHATASE

Elia Lopez, Praveen Kumar Kathare, Thilanka Jayaweera, Prabesh Ghimire, Nicholas Siepert, Damian Raymond, Lauren Minter, Sunethra Dharmasiri and Nihal Dharmasiri

Department of Biology, Texas State University, San Marcos, TX

Auxin is a major hormone that regulates plant growth and development through the modulation of both genomic and non-genomic processes. Genomic processes are regulated through the degradation of a group of transcriptional repressors known as Aux/IAA proteins. These repressors are degraded through the $SCF^{TIR1/AFBs}$ (Skp1, Cullin, F-box) ubiquitin ligases, in which F-box proteins TIR1/AFBs function as nuclear co-receptors of auxin. Recent studies indicate that the function of $SCF^{TIR1/AFBs}$ may also be regulated by other signaling pathways. Through a genetic screen using picloram, a synthetic auxinic herbicide, we isolated Arabidopsis ibr5-4 mutant. IBR5, a dual specificity phosphatase, negatively regulates Aux/IAA degradation through an unknown mechanism. In a yeast two hybrid assay, IBR5 interacts with small GTPases and this interaction occurs through the catalytic domain of IBR5. Interestingly, previous studies have shown that ROP2 and ROP6, two small GTPases, regulate interdigitation of leaf epidermal cells in auxin dependent manner. Similarly, we found that ibr5 mutants are also defective in epidermal cell interdigitation suggesting that these proteins may function in the same pathway. In this report we discuss the possible regulation of plant auxin response by IBR5 through small GTPases.

CHARACTERIZATION OF AN ARABIDOPSIS PICLORAM TRANSPORTER AND ITS TOMATO HOMOLOG FOR DEVELOPING HERBICIDE AND DROUGHT RESISTANCE IN CROP PLANTS

Sunethra Dharmasiri, Praveen Kumar Kathare, Ethan Ginsberg and Nihal Dharmasiri.

Department of Biology, Texas State University, San Marcos, TX

Auxinic hormones play a major role in plant growth and development. Exploiting this phenomenon, auxinic herbicides are widely being used in agriculture and forestry to control invading dicot species. Picloram is a common auxinic herbicide, however, its mechanism of action in plants is not well understood. In a genetic screen for picloram resistant mutants in Arabidopsis, we identified three mutant alleles of PIC30 (PICLORAM RESISTANT 30) gene. PIC30 codes for an MFS major facilitator superfamily (MFS) protein. All three Arabidopsis mutants (pic30-1, pic30-2 and pic30-3) confer selective resistance to picolinate herbicides, and are significantly more resistant to drought. The pic30-3 null mutant is highly defective in transport of picloram, but not the natural auxin IAA, into root cells.

Tomato genome contains three close homologs of PIC30. In an attempt to analyze the role of tomato PIC30 protein, we cloned one of the PIC30 homologs from microtom cultivar of tomato, and introduced the pic30-1 mutation. Overexpression of mutant tomato pic30-1 gene in wild-type microtom confers picloram resistance, while the overexpression of wild-type PIC30 gene confers picloram hypersensitivity in tomato seedlings. Our data suggest the potential use of PIC30 gene for introducing herbicide and drought resistant crop plants, without affecting the normal growth and development.

MANAGING ANTHOCYANIN CONCENTRATION, PROPORTION, ACYLATION AND HYDROXYLATION OF AUTUMN ROYAL TABLE GRAPE WITH PLANT GROWTH REGULATOR AND APPLIED WATER AMOUNT INTERACTIONS

Sonet Van Zyl, S. Kaan Kurtural, G. Dervishian, James A. Kennedy

Department of Viticulture & Enology, California State University, Fresno, CA

Autumn Royal, a black seedless late season table grape variety has a high retail demand with naturally large berries, good flavor and attractive color. One limiting factor for Autumn Royal grape production is proper color development due to its propensity to overcrop. Plant growth regulators are widely used to increase berry size and enhance color development. Research emphasis has recently focused on the concept of water productivity, which is the amount of marketable product produced per unit of water consumed in evapotranspiration (ET_c). Table grape growers tend to over-irrigate to ensure optimum quality grapes. Plant growth regulator treatments included gibberelic acid (GA_3) and Ethrel® applications and a combination of both with an untreated control. Three irrigation treatments were applied: grower control at 120% ETc, early regulated deficit irrigation (RDI) initiated at fruit set irrigated at 80% ET_c until veraison and late RDI initiated at veraison irrigated at 80% of ET_c until harvest. Initial results showed that the combination of irrigation regimes with PGR treatments had no significant effect on berry size (g) or cluster weight (g), but the cumulative 19 lb boxes packed was higher for the early RDI treatment. Total skin anthocyanin content (mg/kg) was significantly higher for the Ethrel® and combined PGR treatments during early RDI. The late RDI treatment showed higher skin anthocyanins for only the combined PGR treatment. Berry firmness was higher for the combined PGR and early RDI treatments. After cold storage trials berry firmness was significantly higher for GA_3 and the combined PGR treatments. The late RDI treatment had the firmest berries of the three irrigation regimes. Preliminary results showed that the RDI early treatment was the better performer, even in drought situations, but must be combined with PGR's. PGR applications are needed in Autumn Royal production and they have an additive effect on yield and berry color. Using PGR's can enhance berry firmness and storage ability. Also, the combined effects of early RDI and PGR applications would increase yield and storage ability and enhance color development. Irrigation savings of 2 acre-inches was achieved with RDI treatments and resulted in full berry color development. This is an ongoing project that presented initial results based on the 2014 growing season.

MOLECULAR BASIS OF JASMONATE ACTION

Qi Tiancong, Song Susheng, Yan Jianbin, Daoxin Xie

School of Life Sciences, Tsinghua University, Beijing, China

Jasmonates have emerged as an essential growth regulator to control diverse plant developmental processes, such as stamen maturation, root growth, anthocyanin accumulation and senescence. Meanwhile, jasmonates have also been recognized as a vital defense signal to activate stress and wound responses and to mediate plant defense against herbivorous insects and necrotrophic pathogens. Upon perception of active molecules of jasmonate, the JA receptor CORONATINE INSENSITIVE1(COI1) targets the Jasmonate-ZIM-domain (JAZ) proteins for ubiquitination through the Skp/Cullin/F-box (SCF)COI1 complex, leading to subsequent degradation of JAZ proteins in the 26S proteasome pathway. A series of JAZ-repressed transcription factors are then released to further modulate downstream gene expression and regulate multiple JA responses. We have used molecular, biochemical and genetic approaches to reveal mechanism for jasmonate action. The recent research progress will be discussed.

A PHYLOGENETIC, EVOLUTIONARY, AND ENZYMATIC ANALYSIS OF THE SITKA SPRUCE FAMILY OF AUXIN CONJUGATE HYDROLASES

James J. Campanella[1], Pedro M. Smith[1], Nadia Zaben[1], Dalma Enriquez[1], John V. Smalley[2], Sarah Hense[3], Sabine Rößler[3], Jutta Ludwig-Müller[3]

[1]Montclair State University, Dept. of Biology, 1 Normal Avenue, Montclair, NJ 07043, USA

[2]Bergen Community College, Dept. of Biology and Horticulture, 400 Paramus Rd., Paramus, NJ 07652, USA

[3]Institut für Botanik, Technische Universität Dresden, Zellescher Weg 20b, 01062 Dresden, Germany

It is thought that all higher plants regulate the balance of auxin, primarily indole acetic acid (IAA), through interactions among de novo synthesis, degradation, efflux, influx, and conjugate synthesis/hydrolysis. Auxin is stored in several conjugated forms that are generally thought not to be active. There are two major types of conjugated molecules: amide-linked IAA bound to one or more amino acids, and the ester-linked form mostly bound to sugar(s). These two types of conjugates are found in varying concentrations among the diverse tissues of gymnosperms and angiosperms. Up to 95% of auxin in a plant is conjugated into these storage forms.

We have examined how the ILR1-like auxin conjugate hydrolase family of genes has functionally evolved in the gymnosperm Sitka spruce (Picea sitchensis). Four orthologues that are homologous to the Arabidopsis thaliana AtIAR3 auxin conjugate amidohydrolase have been isolated and cloned from Sitka spruce (PsIAR31, -32, -33, and -34).

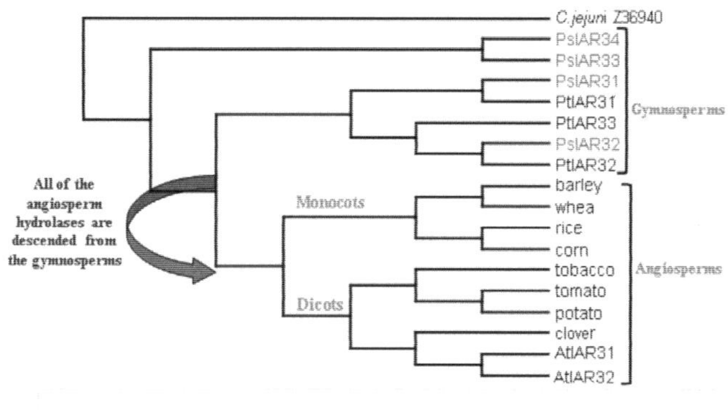

Figure 1. Neighbor-joining cladogram of angiosperm and gymnosperm IAR3 auxin conjugate hydrolase orthologues

Phylogenetic analyses (Fig. 1) were performed to determine whether the gymnosperm hydrolases are evolutionary ancestors of the angiosperm orthologues. The cladogram was rooted with the bacterial hydrolase protein from Campylobacter jejuni (Z36940). PsIAR31, -32, -33, and -34 all branch in clades below the major limb of the bacterial hydrolase. Because of their branch position in the tree, these four spruce proteins appear to be the molecular ancestors of all the other auxin conjugate hydrolases of the separated angiosperm clade (Fig. 1). PsIAR33 and -34 appear to be in the most ancient, ancestral enzyme clade of vascular plants and almost seem to be an outgroup.

Table 1. Enzyme and substrate specificity of Sitka spruce auxin hydrolases. All hydrolysis values are in pmol auxin released/min/ml (average of hree to seven experiments +/- Standard Error)

Enzyme	IAA-Asp	IAA-Ala	IAA-Isoleu	IAA-Phe	IAA-Gly	IAA-Gluc	IAA-Val	IAA-Leu
PsIAR31	0.0	247.0±19.7	249.0±10.7	0.0	0.0	36.7±5.8	0.0	0.0
PsIAR32	482.0±38.5	0.0	0.0	0.0	0.0	96.1±9.3	0.0	0.0
PsIAR33	690.0±303.7	193.3±6.0	42.7±22.6	121.6±32.6	87.7+36.5	412.2±95.8	0.0	441.6±219.6
PsIAR34	580.5±95.8	0.0	456.6±222.9	163.6±134.2	118.3±67.2	0.0	0.0	0.0

The four spruce hydrolases, as a group, recognized a majority of the common auxin conjugates tested as substrates (Table 1), with the exception of IAA-Valine. Uninduced controls of the hydrolase vectors hydrolyzed little or no substrate. PsIAR31 was able to hydrolyze IAA-Ala, IAA-Isoleu, and IAA-Gluc, while PsIAR32 was able to recognize only IAA-Asp and IAA-Gluc as substrates. The more evolutionarily ancient hydrolases, PsIAR33 and -34, recognized a much wider variety of substrates (IAA-Asp, IAA-Ala, IAA-IsoLeu, IAA-Phe, IAA-Gly, and IAA-Leu). Among all the spruce hydrolases the recognized substrates cover the entire gamut of amide and ester bond IAA conjugates found in the more recently evolved angiosperms. There also appears to be functional redundancy in substrate recognition across the various members of the Sitka spruce paralogue family. For example, IAA-Aspartate is recognized by three of four members of the hydrolase family (Table 1). Finally, we have observed the hydrolysis of "longer" auxin conjugates of Indole butyric acid and Indole propionic acid by the spruce conjugate hydrolases. This result suggests that recognition of such conjugates extends back 300-310 million years to the first appearance of gymnosperm.

SOYBEAN (GLYCINE MAX) GROWTH AND YIELD IN RESPONSE TO SEED APPLIED BIOSTIMULANTS IN NEBRASKA

Michael D. Rethwisch, Stephen J. Boggs, Kimberly R. Topil, and Scott Willet.

University of Nebraska-Lincoln, Butler County Extension, 451 N. 5th Street, David City, Nebraska, USA 68632-1666 (402) 367-7410 mrethwisch2@unl.edu

INTRODUCTION

There are several product classes of commercially available chemistries as seed treatments for soybeans. These classes include fungicides, insecticides, and inoculants, as well as biostimulant products. It is not unusual to have multiple chemistry classes applied to soybeans prior to planting, and a number of commercial products contain two or more chemistry classes.

While large data sets exist for many types of products, robust replicated data for biostimulant products obtained under commercial field trials are not as readily available, especially under various field conditions. This project compared inoculants vs. biostimulants, evaluated several common biostimulants applied to soybean seed, compared growth and yield results, and calculated economic values of soybeans treated with biostimulants under irrigated vs. rain-fed conditions in eastern Nebraska.

METHODS and MATERIALS

A series of large scale field experiments were conducted during 2010-2014 using randomized complete block designs and a minimum of four replications. Plots were field length (245-770 meters) and varied in width from 6-14 rows wide with most having 0.76 meter row widths. The majority of experiments were conducted primarily on rain-fed fields, with fields with center pivot irrigation systems being used for irrigated field sites. All fields were in a corn-soybean annual rotation. Soybean fields did not receive additional fertilizer, with the growing crop utilizing available fertility from the previous crop.

Soybean seed was commercially treated with various biostimulants. Two biostimulants (Optimize[7] and BioForge[7]) were used most frequently over this time period, reflecting commercial use and availability, and were used for data comparisons across locations.

Optimize[7] (Novozymes BioSciences, Inc.) contains both an inoculating bacteria (*Bradyrhizobium japonicum*) as well as LCO (Lipochitooligosaccharide) Promoter Technology[7] (minimum of 0.0000002% LCO). The LCO technology is a natural biological signal molecule which advances both root and shoot development, improves plant health, and results in plants that can better withstand environmental stresses.

BioForge[7] (2-0-3 fertilizer and contains N,N'-diformyl urea; Stoller USA) was used at top of label rate of 118 ml/45.5 kg seed. BioForge[7] product labeling notes that it significantly improves plant growth as it up-regulates key genes associated with plant stress, improves root-hair growth to enhance nutrient uptake, and reduces excess ethylene from stress.

Unique field experimentations in addition to regular testing included an inoculant comparison with biostimulant trial, and a trial which examined variety interaction with biostimulants.

Inoculants vs. biostimulants (Waverly site, Lancaster County)

As few replicated large scale field data existed under Nebraska conditions, producers were inquiring if there was a benefit of utilizing biostimulants. In 2010 a field experiment was initiated that evaluated products that contained inoculants and/or biostimulants when applied as seed treatments at top of label directions and compared these with each other and untreated soybeans or soybeans receiving only a fungicide seed treatment.

Nine products were evaluated and compared with untreated soybeans. Treatments included a fungicide only treatment (Trilex^7 2000, active ingredients = metalaxyl and trifloxystrobin (Bayer CropSciences)), various products classified as inoculants which included *Bradyrhizobium japonicum* without a biostimulant (GraphEx SA™, XcitBio^7 SoyRhizo), hybrid (inoculant + biostimulant) products (Trident™, Optimize^7), and products classified as biostimulant only. These latter products were GreenSol™ 48, Heads Up^7, BioForge^7 ST, and ST-822.

GraphEx SA™ (Advanced Biological Marketing) is a specially formulated blend of Excalibre™ soybean inoculant, plus SabrEx™ root inoculant formulated in a talc-graphite carrier. Guaranteed analysis of 5 x 10^9 *Rhizobia* and 1 x 10^8 *Trichoderma* spp. cfu/gm. Use rate was 29.6 ml/22.7 kg soybean (2x recommended regular rate) due to standing water on field for some time prior to planting.

XcitBio^7 SoyRhizo (XcitBio Technologies Inc.) contains advance growth promoting technology (AGPT) in addition to *Bradyrhizobia japonicum.*

Trident™ Bio-Inducer Technology (Precision Laboratories) contains two primary functioning agents, with one being *Bradyrhizobia japonicum* and non-pathogenic *Rhizobium* bacteria, and the other a proprietory blend of nutrients, adhesives and water, which also serve as a bacterium protector.

GreenSol™ 48 (Frit Industries) contains the plant hormones kinetin and gibberellic acid in an 8-20-20 water-soluble fertilizer base (from which the number 48 is arrived). This product is designed to promote plant vigor, early maturity, higher yields and improve crop quality.

Heads Up^7 (marketed by Bayer CropSciences) is an extract from the seed coat of *Chenopodium quinoa.*

ST-822 (Emerald BioAg) is an experimental seed treatment with proprietary ingredients. Little is known about the active ingredient(s) in this product. Usage rate of ST822 was 38 ml/45 kg seed.

Ten plants/plot were dug from the plots in July at six weeks after planting, and nodules were counted and weighed. Root weights were also collected at this time. Grain yields were collected at end of season.

Variety - biostimulant (David City – 2011)

Two soybean varieties (Mycogen 5B261RR, Asgrow 2909) had seed treated with two differing growth enhancement products (ST-822 + the fungicide Trilex[7] 2000, Optimize 400[7]). Seed was treated using a commercial small-scale seed treater on May 20 (13 days prior to planting). The 12 row planter was split, resulting in 6 rows of each variety planted side x side with the same biostimulant in a randomized complete block design for the biostimulant. Optimize[7] and ST822 were used at rates previously stated.

Harvest

Individual plots from each field were combine harvested, with the entire plot being harvested. Grain was then transferred to either a dual axle weigh wagon (+ 0.8 kg), a single axle weigh wagon (+ 8.1 kg) or a grain cart equipped with scales (+ 8.1 kg) depending upon the available equipment. Weight data were recorded. Small sub-samples (approximately 0.25 liter) of harvested soybeans from each plot were collected from the combine, placed in marked re-sealable locking plastic bags, and delivered to the University of Nebraska-Lincoln Department of Agronomy and Horticulture for moisture level determinations. Yield data were then standardized to 13% moisture and yield/plot calculated.

Data analyses

Data were analyzed and treatment means statistically separated using Tukey=s Honestly Significant Difference (HSD) test (JMP 8.0.2 prior to 2013, JMP 10.0.0 thereafter, SAS Institute Incorporated, Cary, NC).

RESULTS

Biostimulants vs. inoculants

No statistical differences existed between any two treatments at this site, however, trends were noted. Significant differences between chemical classes for nodules (nodules/plant and weight of nodules/plant) were noted when product classes were compared with biostimulant only treatments resulting in more nodules and increased nodule weight/plant (Tables 1, 2). While treatments resulted in slightly heavier roots, this was not statistically different (Table 1).

Table 1. Grams/plant of nodules and roots, and nodules/gram of root of Pioneer 93M11 soybeans July 20-24 following planting on June 7, Lancaster County, Nebraska, 2011.

Class	Nodules	Roots	Total	Nodules/g root
No inoculant or biostimulant	0.44 b	3.23a	3.68a	7.1a
Inoculant/Inoculant + Biostimulant	0.56ab	3.30a	3.87a	7.7a
Biostimulant only	0.59a	3.42a	4.01a	8.4a
P Value	*0.02*	*0.76*		*<0.10*

Means in columns followed by the same letter are not statistically different at the p<0.05 level (Tukey-Kramer HSD test, JMP 8.0.2)

Table 2. Mean nodules per Pioneer 93M11 soybean plant, weight (g)/nodule, and percent nodule weight compared with root weight by seed treatment class.

Seed Treatment Classification	Nodules	Grams/nodule	Percent
No inoculant or biostimulant	22.8 b	0.019a	13.8 b
Inoculant/Inoculant + Biostimulant	24.9ab	0.023a	17.4ab
Biostimulant only	28.8a	0.021a	17.7a
P Value	*0.04*	*0.06*	*<0.05*

Means in columns followed by the same letter are not statistically different at the p<0.05 level (Tukey-Kramer HSD test, JMP 8.0.2)

There was a distinct trend for higher yields for products containing a biostimulant, with this averaging 121 kg/ha more than untreated soybeans seed (Table 3), representing an increase of 3.3%.

Table 3. Yield increase of Pioneer 93M11 soybeans in response to various biostimulants applied as seed coating treatment, Lancaster County, NE, 2011.

		Increase over untreated	
Biostimulant treatment	Yield (kg/ha)	Kg/ha	%
BioForge[7]	3,784	139	3.8
HeadsUp[7]	3,792	147	4.0
Optimize[7]	3,766	121	3.3
ST 822	3,775	129	3.6
Trident™	3,716	71	1.9
AVERAGE	**3,767**	**121**	**3.3**

The experiment at David City comparing varietal responses to biostimulants resulted in a consistent yield increase of 0.8-0.9% when ST-822 was applied for both varieties. Results from Optimize[7] were not as consistent with one variety (Asgrow 2909) being documented with a 0.7% increase but the other variety (Mycogen 5B261RR) responding with 131 kg/ha (3.9%) more yield (Table 4). These differences in varietal response are consistent with results from other small plot experimentation involving other varieties and biostimulants in Nebraska, indicating soybean varietal interactions with different biostimulants exist.

When data are compiled for the two biotimulants most often evaluated in large commercial plots in Nebraska (BioForge[7], Optimize[7]), several patterns were able to be noted. Average yield increases over untreated soybeans were noted to be 106 kg/ha, an increase of 3.5% (Table 4).

Usage of the product resulted in a net increase in crop value and net profit (without application cost) with the exception of one field and variety interaction (David City, 2011, Asgrow 2909, Optimize[7]). Average increase was noted to be $25.78/ha for these experiments (Table 4).

Table 4. Soybean yield increases in response to two selected biostimulant seed treatments when produced under rain-fed conditions in eastern Nebraska.

Site and year	Variety	Biostimulant Product	Yield Kg/ha	vs. untreated Kg/ha	%	Cost/ha[1] $ US	Net profit[1]/ha @ $US 33/100 Kg
Waverly 2011	Pioneer 93M11	BioForge	3,784	139	3.8	$9.17	$37.53
"	"	Optimize	3,766	121	3.3	$9.88	$30.17
David City 2011	Asgrow 2909	Optimize	3,126	19	0.7	$9.88	- $ 3.21
" "	Mycogen 5B261RR	Optimize	3,608	131	3.9	$9.88	$33.35
David City 2011	Pioneer 93M11	BioForge	2,897	168	7.1	$9.17	$47.27
" "	" "	Optimize	2,837	120	4.9	$9.88	$29.72
David City 2013	Mycogen 93M11	Optimize	3,733	47	1.3	$9.88	$ 5.63
		AVERAGE		**106**	**3.5**		**$25.78**

[1]*Includes cost of product only, not additional cost of applying product to seed*

Data from center pivot irrigated fields, while experimentation has not yet been as robust as that from rain-fed fields, indicate positive results. Biostimulants applied as seed treatments in soybeans have thus far been recorded to increase grain yield by an average of 36 kg/ha, an average of 0.7% more than untreated soybeans (Table 5).

This increase is positive relative to net value/ha ($5.19), however the price of applying the product to the seed has not been included and would be expected to reduce this value (Table 5).

Table 5. Soybean yield increases in response to biostimulant seed treatment when produced with center pivot irrigation in Nebraska.

Site and year	Variety	Biostimulant Product	Yield Kg/ha	vs. untreated Kg/ha	%	Cost/ha[1] $ US	Net profit[1]/ha @ $US 33/100 Kg
Rising City 2009	NK S30-D4	BioForge	4,836	+44	0.9	$9.17	$4.18
David City 2014	Seitec 8261RR	PreCede	4,900	+29	0.6	$3.41	$6.20
		AVERAGE		**+36**	**0.7**		**$5.19**

[1]*Includes cost of product only, not additional cost of applying product to seed*

Average yield increases noted thus far in Nebraska from biostimulants have been greater in rain-fed production systems than in center pivot irrigated systems (Table 6), with the average difference being 70 kg/ha (2.8%). While the exact reason for this is not fully known, it is thought that irrigated soybeans would have less moisture stress, and irrigation waters in Nebraska also contain nitrates which are utilized by growing crops.

Table 6. Comparison of irrigated vs rainfed average field yield increases and values

	Increases vs. untreated		Profit[1]/ha @
Field type	Kg/ha	%	US $33/100kg
Rain-fed fields	106	3.5	25.78
Irrigated Fields	36	0.7	5.19

[1]Includes cost of product only, not cost of applying product to seed

GIBBERELLIN TREATMENT TO A SELECTION OF COMMERCIAL CULTIVARS OF THE GARDEN PEA (PISUM SATIVUM L.)

Sonja L. Maki

Department of Plant and Earth Science, University of Wisconsin-River Falls, River Falls, Wisconsin, USA

Gibberellins regulate several different aspects of pea growth and development, such as plant height, axillary branch development, and floral characteristics. In a previous study, gibberellin treatment improved yield in a determinate, multiflowered line of pea (Maki, et al, 2005). In the current study, three commercial cultivars of the garden pea were treated with different gibberellins to determine their effect on reproductive development. The cultivars 'Super Sugar Snap,' 'Gonzo', and 'Petite-Snap Greens' were treated with different gibberellins when plants had four expanded leaves. Results of the treatments on plant height, tendril development and floral development of these commercial cultivars will be discussed. One interesting observation was doubled petal and multiple carpel production in the 'Petite-Snap Green' cultivar.

Maki, S.L., Mullen, H., Pharis, R. and Singer, S. 2005. Improvement of Yield in Greenhouse Grown Determinate Multiflowered Peas with Gibberellin Treatments. Proceedings of the 32nd Annual Conference of the Plant Growth Regulation Society of America, Newport Beach, California.

META-ANALYSIS OF PERFORMANCE OF ASCEND® PLANT GROWTH REGULATOR IN ANSWER PLOT® TESTING SYSTEM

Stephanie Wedryk and Catherine White

WinField Solutions, LLC, Saint Paul, MN, USA

Plant growth regulators are an important tool for in-season management of agronomic row crops to achieve maximum yield potential. For early season management of corn, in-furrow application of plant growth regulators can enhance germination, root growth, and biomass resulting in higher grain yields. Testing plant growth regulators under conditions that closely mimic a grower's fields is crucial for generating data that can support their use by growers. However, determining the effect of in-furrow applied plant growth regulators on grain yield is complicated by planting conditions, events during the growing season, and the inherent variability in field research. To account for factors that influence plant growth regulator performance and data, a meta-analysis was performed to determine the mean yield response to and odds of yield improvement by Ascend® plant growth regulator. From 2011 to 2013, Ascend® plant growth regulator, a combination of gibberellic acid, kinetin, and indole butyric acid, was tested at 251 site-years across the Midwestern and Southern United States. Mean yield was different for Ascend® plant growth regulator in each year of testing ($P = 0.007$). However, there were no differences in mean yield due to soil temperature at planting, precipitation, or test site location ($P > 0.05$). The odds ratio indicated improved odds of yield increase with Ascend® plant growth regulator with differences in performance due to testing year, geography and soil temperatures ($P < 0.05$). The odds ratio indicated higher odds of yield increase due to Ascend® plant growth regulator when soil temperatures at planting were less than 10°C (odds ratio = 4.2) compared to 10-15.6°C (odds ratio = 2.9) or greater than 15.6°C (odds ratio = 1.6). Meta-analysis can be used to identify the circumstances under which the greatest chance of yield improvement with plant growth regulators exist to help growers make decisions for product applications.

EFFICACY OF PROMALIN ON THE GERMINATION OF PASSIFLORA ALATA CURTIS SEEDS

Priscila F. Medina[1], Laura M. M. Meletti[1], João J. D. Parisi[1], Camilo L.Medina[2]

[1] *Campinas Agronomic Institute, Campinas, SP, Brazil, 13075-630*

[2] *Conplant- Consulting, Training, Research and Agricultural Development, Campinas, SP, Brazil, 13084-200*

ABSTRACT

The sweet passion-fruit plant (*Passiflora alata* Curtis) is grown in Brazil for fruit production and also for medicinal purposes as it contains passiflorin, a natural sedative. It is propagated by seeds, which manifest the maximum germination, at moisture contents between 18 and 20%, obtained when they are just removed from the fruit. At moisture contents under 13%, the formation of normal seedlings can be reduced, and the seeds lose viability after the first month of storage. The period of time for completing germination of a seed sample of this species is 30 days, favoring the growth of fungi which can infect seeds and seedlings. In order to obtain information on the germination of seeds of *P. alata* exposed to drying or low temperature and to investigate the possibility of improving germination performance of the seeds through the use of plant growth regulators, germination tests were carried out with 19% moisture content seeds, maintained or not at 1 °C for 24 hours and seeds dried at 28 °C until reaching moisture contents of 13.8, 12.1, 9.0 and 8.8%. The seeds of all drying levels were still submitted or not to fungicide treatment with carbendazim + thiram (80 and 200 mL of cp / 100kg of seeds) or to soaking in water or in Promalin solution (1.8% of GA_{4+7} and 1.8% of N- (6-phenylmethyl-1H-aminopurine), at a concentration of 250 mg L^{-1}) for 16 hours and further fungicide treatment. It was noted mainly that the use of Promalin doubled the speed of seed germination, standardized the seedlings growth at all drying treatments and also increased the percentage of primary root protrusion of the 8.8% moisture content dried seeds, but limited the growth of the primary roots and inhibited the formation of lateral roots and root hairs, preventing the subsequent seedling establishment in plant substrates. The role of Promalin on germination and on initial seedling development is discussed.

INTRODUCTION

The *Passiflora alata* Curtis species belongs to the Passifloraceae family and Passiflorales order (Silva and São José, 1994). It is native from South America and occurs in Brazil, Peru, Paraguay and Argentina (Killip, 1938, cited by Ferrari, 2009). Brazilian ecological conditions are favorable for its development and production (Oliveira, 2003).

The interest in cultivating *Passiflora alata* Curtis has grown due to the quality of the fruit for consumption *in natura* and to the prices achieved on the fresh fruits market, in addition to the high price of the juice on the national and international market. It is also cultivated for medicinal purposes, due to produce passiflorin, a natural sedative (Meletti et al., 2002).

The predominant method for cultivating passion fruit is by planting seedlings originated from seeds (Wagner Junior et al., 2007). However, there are many reports of low percentage of germination of *P.alata* seeds, which hinders the formation of vigorous seedlings and rootstocks. However, these seeds are not considered as recalcitrant, because there is information about seeds of this species that have preserved their germination potential unchanged for 168 days, when stored with moisture content between 5 and 10% (Osipi and Nakagawa, 2005).

Growth regulators such as gibberellins (gibberellic acid, GA_3, GA_4 and GA_7), cytokinins (mainly the kinetin and benzyladenine) and ethylene are related to the overcoming of seed dormancy. These substances can improve the percentage, uniformity and speed of germination, resulting in production of more vigorous seedlings (Taiz and Zeiger, 2009).

Gibberellins act in the synthesis of germination specific proteins and RNA, for overcoming dormancy and controlling the hydrolysis of reserves, which are degraded into starch and protein used for developing the embryo and to promote cell elongation, leading to the shattering of the tissues that restrict the primary root growth, as the endosperm, the seed coat or structures of the fruit (Taiz and Zeiger, 2009).

Cytokinins also act on seed germination (Mayer and Poljakoff-Mayber, 1978), regulating the level of inhibitors present in the seeds, allowing them to become more sensitive to the action of gibberellin (Walker et al., 1989). Gepstein and Ilan (1980) demonstrated that the cytokinins induce the increase of the activity of hydrolases, as α-amylase and proteases.

Among the effects of using growth regulators are the increase of endogenous level of these compounds or antagonizing the effect of inhibitors on embryonic metabolism (Zaidan and Barbedo, 2004).

The promotion of the germination process was not achieved in some studies on the application of plant growth regulators in seeds of *P. alata,*, as the ones realized by Coneglian et al. (2000), using gibberellic acid at concentrations of 0, 100 e 300 mg L^{-1}; and by Alves et al. (2006) using gibberellin (75 e 1000 mg L^{-1}), cytokinin (250 and 750 mg L^{-1}) and KNO3 (0.1 and 0.2 %).

In other studies carried out with gibberellins it was noted an increase in the percentage of *P. alata* seeds germination, when immersed in 100 mg L^{-1} of GA3 (Ferreira,1998), or when submitted to pre-soaking in contact with substrate moistened with solutions of 150 and 300 mg L^{-1} of gibberellic acid (Rossetto et al., 2000), or as soaked in solutions of GA_3, at the concentration of 500 mg L^{-1}, which promoted the greater level of responses, without interference of the times of soaking (Ferreira et al., 2001b) and when immersed in gibberellic acid solutions for five hours, 100, 500, 750 and 1000 mg L^{-1} (Ferreira et al, 2001a).

However the application of gibberellins associated with cytokinin (GA4+7 + N-(phenylmethyl)-1H-purine-6-amine.), at concentrations of 200 and 250 mg L^{-1}, proved to be even more beneficial to *P. alata* Curtis seeds, leading to an increase in the percentage and speed of germination and to a reduction of the time for completing germination and of the dormant seeds percentage (Ferrari et al,. 2008 and Ferrari, 2009). It has been also noticed that seeds of this species soaked in gibberellins associated with cytokinin present an increase of total amylase and when treated with a 100 mg. L^{-1} GA4+7 + N-(phenylmethyl)-1H-purine-6-amine solution its metabolic activity was accelerated during the phase II of the germination, resulting in faster growth of the embryo and in anticipation of the protrusion of the primary root (Ferrari, 2009).

Considering that a) sweet passion-fruit seeds manifest the maximum germination, at moisture contents between 18 and 20, just after being removed from the ripe fruit; b) the formation of normal seedlings can be reduced at moisture contents of the seeds under 13%; and these seeds often lose viability after one-month storage in environmental conditions; c) the period of 30 days is required for germinating a sample of these seeds, favoring fungi growth, which can infect seeds and seedlings, this research was carried out in order to obtain information on the germination of seeds of *P. alata* exposed to drying or low temperature and to investigate the possibility of improving germination performance of the seeds through the use of Promalin (GA4+7 + N-(phenylmethyl)-1H-purine-6-amine.).

MATERIALS AND METHODS

Seeds of sweet passion fruit were obtained from ripe commercial fruits. The aryl was removed and the seeds with initial moisture content of 19.1% were dried inside a forced air oven at 30 °C. The following treatments were performed: **D1** – without drying and moisture content (MC) of 19.1%; **D2** – without drying, kept at 1 °C for 24 hours (18.3% MC); **D3** – dried for 1,5 hours (13.8% MC); **D4** – dried for 4 hours (12.1 % MC); **D5** – dried for 25 hours (9.0% MC); **D6** – dried for 40 hours (8.8% MC).

Before beginning drying seeds were weighed on a digital balance with precision of 0.0001 g, and also at one-hour intervals, during the drying process, until reaching the weight correspondent to the desired moisture content.

To determine the moisture content (MC) of the seeds before and after drying, two replications of 25 whole seeds were weighed (W1) on a digital balance with precision of 0.0001 g, taken to an oven at a temperature of 105 ºC for 24 hours and reweighed (W2). In this procedure, in addition to moisture content, based on the equation MC = W1-W2/W1, the dry matter (W2) determined before drying was used for calculating the final weigh of the samples that should be dried until reaching each desired moisture content level.

Seeds of all drying levels were submitted to four treatments: **T1** – Control; **T2** – fungicide: carbendazim + thiram (80 and 200 mL of cp / 100kg of seeds); **T3** – immersion in distilled water oxygenated by aquarium air pump for 16 hours and trestment with the same fungicide used in T2; **T4** – immersion in Promalin (1.8% of GA4 + 7 and 1.8% of N- (6-phenylmethyl-1H-aminopurine) solution - 250 mg L^{-1} oxygenated by aquarium air pump for 16 hours and treatment with the same fungicide used in T2.

The germination test was conducted with four replications of 25 seeds of all treatments, distributed on paper towel moistened with distilled water at a ratio of 2.5 times the weight of the dry paper (Brasil, 2009). Sowing was performed on two overlaid sheets and one used as covering, making up rolls which were maintained in a germination chamber (B.O.D. - Biochemical Oxygen Demand) under controlled alternating temperature of 20-30 ºC. The substrates were rewetted as necessary. Primary root protrusion was evaluated daily, up to the 40th day. Normal seedlings were counted up to the 14th and 40th days. Dormant seeds were registered up to the end of the test. The results were expressed in percentage. The germination (primary root protrusion) speed index was also determined (Maguire, 1962).

Nineteen days after seeding, hypocotyls, primary roots and seedlings were measured in millimeters. Obtaining the average length of hypocotyl, primary root and seedling of the plot followed the AOSA (Association of Official Seed Analysts, 2009) recommendation: the sum of the lengths, divided by the number of seedlings measured. Once the measures were taken, the hypocotyls-primary root axes were separated from the cotyledons and taken to the oven at a temperature of 80± ºC for 24 hours. After this period the material was weighed, giving rise to the dry matter of the experimental plot. This value was divided by the number of seedlings placed to dry, resulting in dry matter weight per individual.

The experiment design was completely randomized with for replications of 25 seeds. The data obtained were subjected to the variance analysis (F test), at 5% level, using the SISVAR computer system (Ferreira, 2011). When relevant, the averages were compared by the Tukey test, also at 5% (Santana and Ranal, 2004).

RESULTS AND DISCUSSION

Germination results of seeds of *P. Alata* are shown in Figure 1. Dormancy (Figure 1C) increased only as untreated seeds (T1) were dried for a longer period of time (D6) and the 1°C-temperature supplied for 24 hours (T2) did not induce secondary dormancy on seeds newly removed from ripe fruits. It was also verified a trend toward an increase of dormancy incidence due to the fungicide (carbendazim + thiram) treatments (T2 and T3); except for the most dried seeds (D5 and D6). On the other hand, seeds immersed in gibberellins associated with cytokinin (GA4+7 + N-(phenylmethyl)-1H-purine-6-amine.) solution, for 16 hours, did not became dormant, even after being treated with fungicide (T4), confirming the results obtained by Ferrari et al. (2008) and Ferrari (2009). This treatment (T4) also significantly increased the speed of germination of seeds of all drying treatments (Figure 1B), while the immersion in distilled water for the same period (T3) was detrimental to the speed of germination, although the oxygen supplement, leading to lower growth of seedlings, measured 19 days after the installation of the germination test (Figure 4A, B, C and D).

The beneficial effect of the association of GA4+7 + N-(phenylmethyl)-1H-purine-6-amine.on the germination of seeds of *Passiflora alata* Curtis were also observed by Ferrari et al. (2008) and Ferrari (2009). Gibberellins and cytokinins are compounds related to dormancy overcoming and when supplemented, increase the endogenous level of these compounds or antagonize the effect of inhibitors on embryonic metabolism and can also accelerate the germination of not dormant seeds (Zaidan and Barbedo, 2004).

Gibberellins are known as promoting the production and/or secretion of several hydrolytic enzymes during germination, among which is the α-amylase, involved in dissolving the reserves of the endosperm, resulting in sugars, amino acids and other products transported to the embryo, which are needed for its growth (Taiz and Zeiger, 2009). The exogenous application of giberellins influence protein metabolism and could double the rate of seeds protein synthesis (Mc Donald and Khan, 1983).

Cytokinins are also involved in the germination process, complementing the action of gibberellins in the induction of germination and enzymatic processes, as these events are blocked by inhibitors (Fraga, 1982), in addition to increasing cell division (Taiz and Zeiger, 2009) and inducing increases in the activity of α-amylase and proteases (Gepstein and Ilan, 1980).

Ferrari (2009) confirmed that when exogenous gibberellins are associated with cytokinin (GA4+7 + N-(phenylmethyl)-1H-purine-6-amine.) and supplied to *Passiflora alata* seeds, GA4 and GA7 are transformed to GA1, promoting an increase in the total amylase activity, increasing the reduction of total soluble sugars levels and the degradation of these sugars, with the formation of reducers soluble sugars, which are used as a source of energy for the growth of the embryo in phase II of germination and the completion of this process in the beginning of phase III, when the emergence of primary root occurs, as it is described by Castro et al., (2004). These actions of GA4+7 + N-(phenylmethyl)-1H-purine-6-amine result in greater production of vegetable tissue, anticipating the start of phase III of germination.

This explains why the use of Promalin increased the speed of germination and eliminated de incidence of dormancy of sweet passion-fruit seeds inn this research, as it was verified by

Ferrari et al.(2008) and Ferrari (2009), in all drying levels and also why the percentage of germination (Figure 1C) of the most dried seeds was less reduced, as treated with this plant regulator (T4), in comparison to the control seeds (T1).

Nevertheless, this seed treatment limited the growth of primary roots, inhibited the formation of lateral roots and root hairs of seedlings from all the other treatments, preventing the subsequent seedling establishment in plant substrates (Figure 3). Surprisingly, normal seedlings (Figure 2) could be obtained for the most dried seeds (D6), when treated with Promalin (Figure 1D). However, even for these seedlings considered as normal, it was verified a disproportional hypocotyl growth (Figure 4A) in relation to the root growth (Figure 4B), resulting in significantly longer seedlings (Figure 4C) than in the other treatments (T1, T2 and T3). Differences among treatments and drying levels observed in dry mass (Figure 4D) were less significant than those observed in the dimensions measures.

Although cytokinins associated to gibberellins act as promoter of seed germination and induce cell division in early seedling development, resulting in elongation of the primary root, depending on the cytokinins concentration, the formation of lateral roots can be inhibited, the growth of the primary root limited and a disproportionate growth of the hypocotyl is promoted, as it was verified in this study.

It is widely known that both auxins and cytokinins, when applied at supra-optimal concentrations can inhibit the growth of plant organs. In both cases, most of this inhibitory effect is mediated by the induction of ethylene production by the action of ACC synthase enzyme, which catalyzes essential steps in the biosynthesis of this phytohormone. In the specific case of cytokinins, this seems to be one of the main causes of the strong inhibition caused on the root elongation (Peres and Kerbauy, 2012).

Therefore, although the method of soaking *P. alata* seeds for 16 hours in a 250mg.L^{-1} GA4+7 + N-(phenylmethyl)-1H-purine-6-amine solution has promoted seed germination, confirming the results obtained by Ferrari et al. (2008) and Ferrari (2009), this procedure may have provided a supra-optimal concentration of cytokinin, for the seedlings of *P. Alata* development, generating a relative proportion auxin/cytokinin favorable to cytokinins, which as described by Skoog and Miller (1957) cited by Pere and Kerbauy (2012), could induce the formation of shoot both *in vitro* and *ex vitro*, which may explain the disproportion between the development of hypocotyl and root in this research.

Considering that hormonal responses differ between tissues of germinating seeds and vegetative tissues and the few studies carried out on hormonal control in growing seedlings, researches on this subject are still required in order to improve the establishment of *P. alata* seedlings.

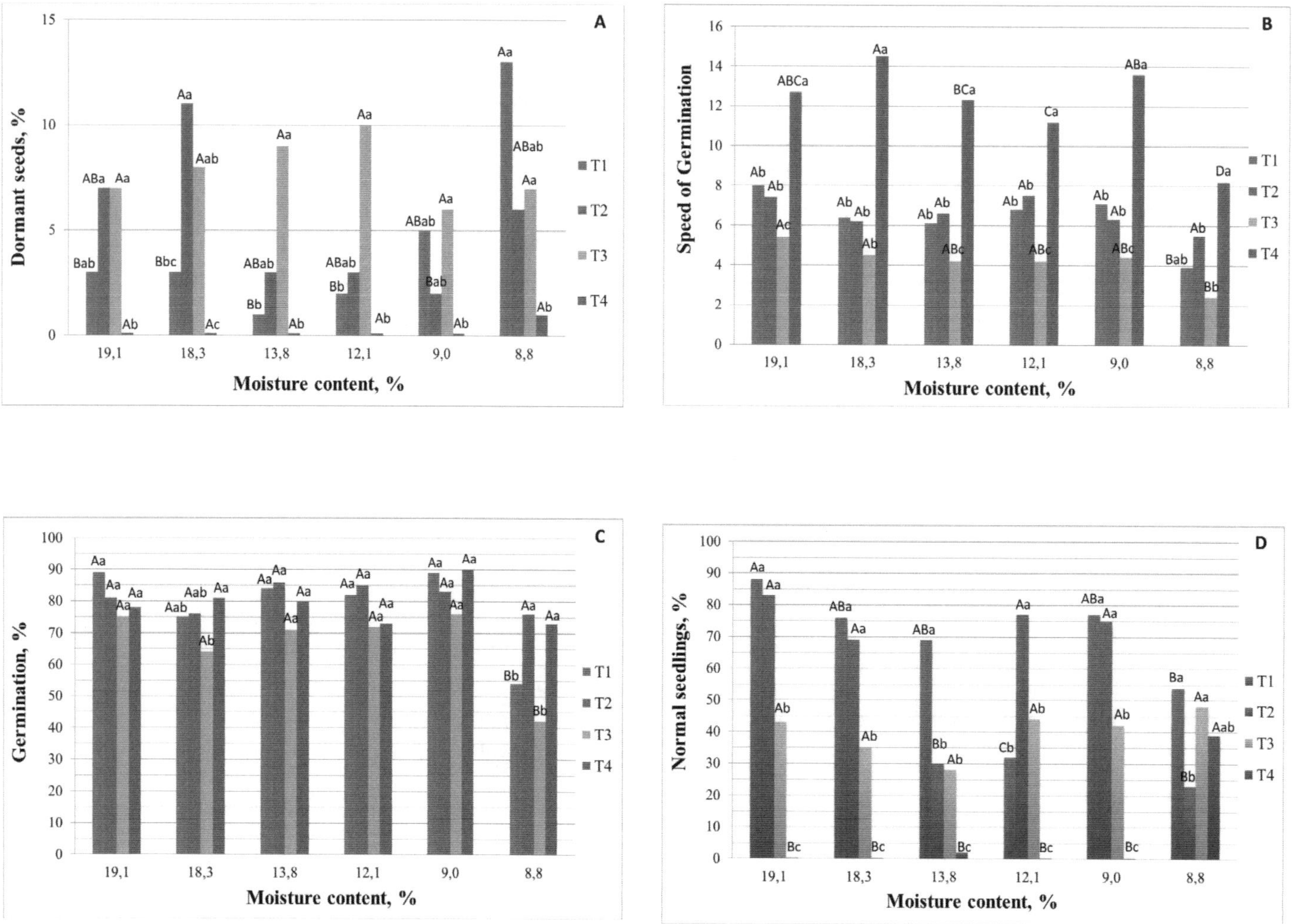

Figure 1. Mean values of *P. alata* dormant seeds (**A**), index of speed of germination (**B**), germination (**C**) and normal seedlings (**D**), according to the seeds moisture content and seed treatments: control (**T1**), fungicide carbendazim + thiram - 80 and 200 mL of cp / 100kg of seeds (**T2**); immersion in distilled water and treated with the fungicide (**T3**) immersion in Promalin (1.8% of GA4 + 7 and 1.8% of N- (6-phenylmethyl-1H-aminopurine) solution - 250 mg L^{-1} and treated with the fungicide (**T4**). Columns with the same letter (uppercase comparing moisture content, lowercase comparing treatments) do not differ among them (Tukey, 5%).

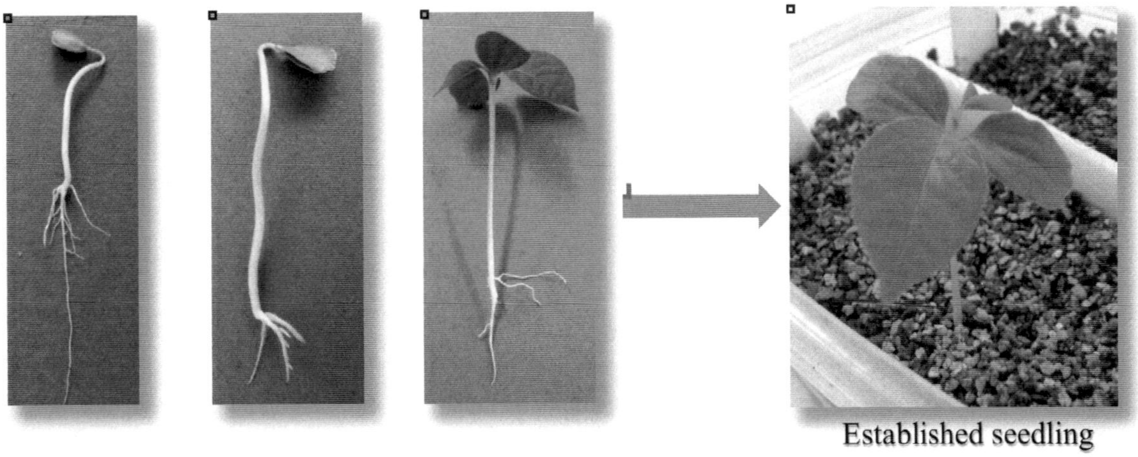

Figure 2: Seedlings considered as normal, by having all the necessary structure for becoming adult plants.

Figure 3: Seedlings considered as abnormal, by having not all the necessary structure for becoming adult plants.

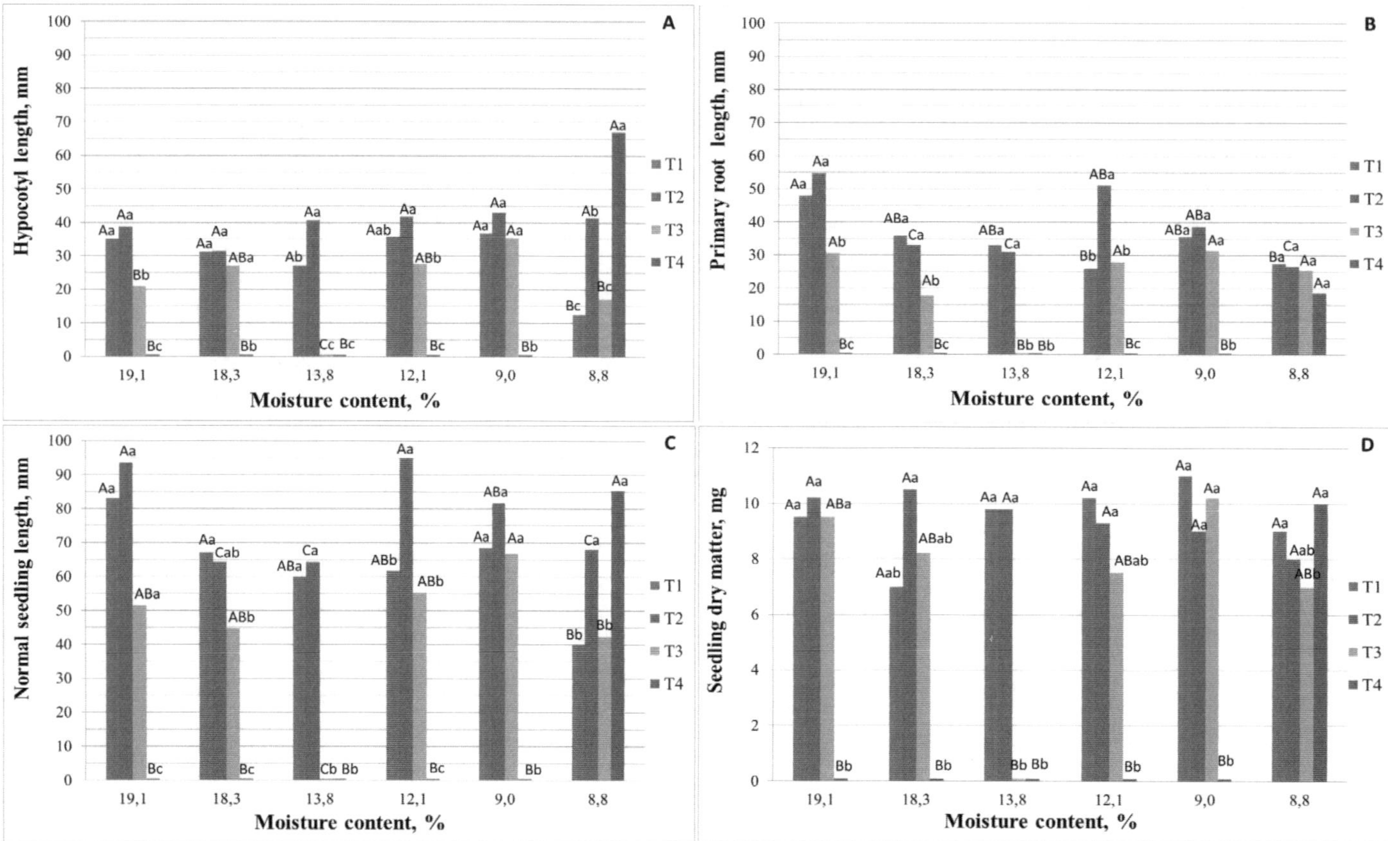

Figure 4. Mean values of hypocotyl length (**A**), primary root lenght (**B**) total length (**C**) and dry matter (**D**) of *P. alata* normal seedlings, according to the seeds moisture content and seed treatments: control (**T1**), fungicide carbendazim + thiram - 80 and 200 mL of cp / 100kg of seeds (**T2**); immersion in distilled water and treated with the fungicide (**T3**) immersion in Promalin (1.8% of GA4 + 7 and 1.8% of N- (6-phenylmethyl-1H-aminopurine) solution - 250 mg L^{-1} and treated with the fungicide (**T4**). Columns with the same letter (uppercase comparing moisture content, lowercase comparing treatments) do not differ among them (Tukey, 5%).

CONCLUSIONS

- Secondary dormancy can be induced in seeds of *P. alata*, by drying them at 30°C for at least 40 hours, until they reach moisture contents of 8.8 %.
- The exposure to a low temperature (1°C), for 24 hours, does not induce secondary dormancy in seeds of *P. alata*, newly removed from ripe fruits.
- The immersion of seeds of *P. alata* in a solution of Promalin (1.8% of GA4 + 7 and 1.8% of N- (6-phenylmethyl-1H-aminopurine), at a concentration of 250 mg.L^{-1}, for 16 hours, with supplementation of O_2, increases the speed of germination, but limits the growth of the primary root and inhibits the formation of lateral roots and root hairs and promotes the elongation of the hypocotyl, originating abnormal seedlings.

LITERATURE CITED

ALVES, C.Z.; SÁ, M.E.; CORRÊA L.S.; BINOTTI, F.F.S. Efeito da temperatura de armazenamento e de fitorreguladores na germinação de sementes de maracujá-doce e desenvolvimento inicial de mudas. *Acta Scientiarum Agronomy*, v.28, p 441-448, 2006.

AOSA. ASSOCIATION OF OFFICIAL SEED ANALYSTS. *Seed Vigor Testing Handbook*, 2009, 341p.

BRASIL. Ministério da Agricultura, Pecuária e Abastecimento. *Regras para Análise de Sementes.* Ministério da Agricultura, Pecuária e Abastecimento. Secretaria de Defesa Agropecuária. Brasília: MAPA/ACS, 2009. 395p. http://www.agricultura. gov.br/arq_editor/ file/2946_regras_analise _sementes.pdf

CONEGLIAN, R.C.C.; ROSSETTO, C.A.V.; SHIMIZU, M.K.; VASCONCELLOS, M.A.S. Efeitos de métodos de extração e de ácido giberélico na qualidade de sementes de maracujá-doce (*Passiflora alata* Dryander). *Revista Brasileira de Fruticultura*, v. 22, n.3, p.463-467, 2000.

FERRARI, T. B.; FERREIRA, G.; MISCHAN, M.M.; PINHO, S.Z. Sweet passion fruit (*Passiflora alata* Curtis) germination: phases and effect of plant growth regulator. *Biotemas,* v.21, n.3, p.65-74, 2008. https://periodicos.ufsc.br/index.php/biotemas/article/view/2175-7925.2008 v21n3p65

FERRARI, T.B. *Metabolismo germinativo de sementes de Passiflora alata Curtis tratadas com giberelinas e citocinina.* Tese de Doutorado, Faculdade de Ciências Agronômicas, Universidade Estadual Paulista, Botucatu, 103 p. 2009.

FERREIRA, D.F. sisvar: a computer statistical analysis system. *Ciência e Agrotecnologia*, v.35, n.6, p.1039-1042, 2011. http://www.scielo.br/pdf/cagro/v35n6.pdf

FERREIRA, G. *Estudo da embebição e do efeito de fitorreguladores na germinação de sementes de Passifloráceas.* 1998. 144p. Tese (Doutorado em Horticultura), Faculdade de Ciências Agronômicas, Universidade Estadual Paulista, Botucatu, 1998.

FERREIRA, G.; FOGAÇA, L.A.; BLOEDORN, M. Efeito do ácido giberélico (GA_3) aplicado em sementes de maracujá-doce (Passiflora alata Dryander) para a produção de mudas em diferentes embalagens. *Revista Brasileira de Fruticultura*, v. 23, p. 152-155, 2001a.

FERREIRA, G.; FOGAÇA, L. A.; MORO, E. Germinação de sementes de *Passiflora alata* Dryander (maracujá doce) submetidas a diferentes tempos de embebição e concentrações de ácido giberélico. *Revista Brasileira de Fruticultura*, v.23, n.1, p.160-163, 2001b.

FRAGA, A.C. Dormência de sementes. *Informe Agropecuário*, v.8, p. 62-64, 1982. GEPSTAIN S, ILAN I. Evidence for the involvement of cytokinins in the regulation of proteolytic activity in cotyledons of germinating beans. *Plant & Cell Physiology*, v.21, p.57-63, 1980.

MAGUIRE, J.B. Speed of germination-aid in selection and evaluation for seedling emergence vigor. *Crop science*, v.2, n.2, p.176-177, 1962. https://www.crops.org/publications/cs/abstracts/2/2/CS0020020176

MAYER, A.M.; POLJAKOFF-MAYBER, A. The germination of plants. 4.ed. Oxford: Pergamon Press, 1978. 270p.

Mc DONALD, M.D.; KHAN, A.A. Acid scarification and protein synthesis during seed germination. *Agronomy Journal*, v.2, p.111-114, 1983.

MELETTI, L.M.M.; FURLANI, P.R.; ÁLVARES, V.; SOARES-SCOTT, M.D.;BERNACCI, L.C.; AZEVEDO FILHO, J.A. Novas tecnologias melhoram a *produção* de *mudas* de *maracujá. O Agronômico*, Campinas, *v.54*, n.1, p.*30-33, 2002.*

OLIVEIRA, A. *Uso de reguladores vegetais na formação de mudas de Passiflora alataDryander.* 2003. 101p. Dissertação (Mestrado), Centro de Ciências Agrárias, Universidade Estadual do Oeste do Paraná, Marechal Cândido Rondon, 2003.

OSIPI, E. A. F.; NAKAGAWA, J. Avaliação da potencialidade fisiológica de sementes de maracujá-doce (*Passiflora alata* Dryander) submetidas ao armazenamento. *Revista Brasileira de Fruticultura*, v. 27, n. 1, p. 52-54, 2005.

PERES, L.E.P.; KERBAUY, G.B. Citocininas. In: KERBAUY, G.B. ed. *Fisiologia Vegetal.* 2ed. Rio de Janeiro: Guanabara Koogan, p.212-234, 2012.

ROSSETTO, C.A.V.; CONEGLIAN, R.C.C.; NAKAGAWA, J.; SHIMIZU, M.K.; MARIN, V.A. Germinação de sementes de maracujá-doce (*Passiflora alata* Dryand) em função de tratamento pré-germinativo. *Revista Brasileira de Sementes*, v.22, n.1, p.247-252, 2000.

SANTANA, D.G.; RANAL, M.A. *Análise da germinação*: um enfoque estatístico. Brasília: Universidade de Brasília, 2004. 248p.

SILVA, A.C. da; SÃO JOSÉ, A.R. Classificação Botânica do Maracujazeiro. In: SÃO JOSÉ AR, ed. *Maracujá: produção e mercado.* Vitória da Conquista: DFZ/UESB, p. 1-5, 1994.

TAIZ, L.; ZEIGER, E. *Fisiologia vegetal. 4.ed. Porto Alegre: Artmed, 2009.* 819p.

WAGNER JÚNIOR, A.; NEGREIROS, J.R.S.; ALEXANDRE, R.S.; PIMENTEL, L.D.; BRUCKNER, C.H. Efeito do pH da água de embebição e do trincamento das sementes de maracujazeiro amarelo na germinação e desenvolvimento inicial. *Ciência e Agrotecnologia* v.31, p. 1014-1019, 2007.

WALKER, M.A.; ROBERTS, D.R.; WAITE, J.L.; DUMBROFF, E.B. Relationship among cytokynin, ethylene and polyamines during the stratification-germination process in seeds of *Acer saccharum*. *Physiologia Plantarum,* v. 76, p. 326-332, 1989.

ZAIDAN, L.B.P.; BARBEDO, C.J. Quebra de dormência em sementes. In: Ferreira, A.G.; BORGHETTI, F. eds. *Germinação: do básico ao aplicado.* Porto Alegre: Artmed, 135-162, 2004.

APPLICATIONS OF STIMPLEX™, A COMMERCIAL BIOSTIMULANT, IMPROVES RED COLOR ON 'JONAGOLD' APPLES

Holly Little

Acadian Seaplants

Peel color in apples is an important quality attribute, and is important in determining market acceptance. Red color development is dependent on environmental conditions, and many different cultural techniques have been used in an attempt to try to improve red color development.

Field trials in Wayne County, NY were conducted between 2012 and 2014 to evaluate the effect of Stimplex™ (*Ascophyllum nodosum* extract, Acadian Seaplants) on red color development in 'Jonnagold' apples. A full season Stimplex program was compared to a grower standard program, which included an additional biostimulant. Percent red color was highest in 2014, and lowest in 2012. In 2012 and 2014, the percent red fruit color was significantly higher in fruit from Stimplex treated trees than in the grower standard (65% vs 71.4%, and 78% vs 84%). In 2013, red color in Stimplex treated fruit was numerically higher than the control (72.7% vs. 76.3%). In 2013 and 2014, the same trees received the same treatments, but pre-harvest plots were split, and half the plots received a preharvest treatment of Blush (prohydrojasmon, Fine Americas, Inc). Percent red color with Blush was the same as the Stimplex treatment, but significantly greater than the grower standard control. In both years, the percent red color in the Stimplex plus Blush treatment was significantly greater than the other treatments (2013 72.7% vs 85.4%, 2014 78% vs 91.1%).

These results indicate the benefits of a Stimplex program, under commercial production practices, for improving peel color in Jonagold apples with or without the application of Blush.

RICE RAB11 IS REQUIRED FOR RESISTANCE TO PATHOGENS THROUGH REGULATING JASMONIC ACID SYNTHESIS

Hee Rang Yun and Jae Bok Heo

Department of Molecular Biotechnology, Dong-A University, Busan, Korea

Rab GTPases are a large family of proteins with a variety of regulatory functions in membrane traffic and development. Previously, we characterized OsRab11, which in concert with OsGAP1 and OsGDI3 regulates vesicular trafficking from the trans-Golgi network (TGN) to the plasma membrane or vacuole. To further elucidate the physiological function of OsRab11 in plants, we performed yeast two-hybrid screens using OsRab11 as bait. OsOPR8 was isolated and shown to interact with OsRab11. A co-immunoprecipitation assay confirmed this interaction. The green fluorescent protein-OsOPR8 fusion product was targeted to the cytoplasm and peroxisomes of protoplasts from *Arabidopsis thaliana*. OsOPR8 exhibited NADPH-dependent reduction activity when 2-cyclohexen-1-one (CyHE) and 12-oxo-phytodienoic acid (OPDA) were supplied as possible substrates. Interestingly, NADPH oxidation by OsOPR8 was increased when wild-type OsRab11 or the constitutively active form of OsRab11 (Q78L) were included in the reaction mix, but not when the dominant negative form of OsRab11 (S28N) was included. OsRab11 was expressed broadly in plants and both OsRab11 and OsOPR8 were induced by jasmonic acid (JA) and elicitor treatments. Overexpressed OsRab11 transgenic plants showed resistance to pathogens through induced expression of JA-responsive genes. In conclusion, OsRab11 may be required for JA-mediated defense signaling by activating the reducing activity of OsOPR8.

EVALUATION OF PLANT GROWTH RETARDANTS ON PENSTEMON X MEXICALI 'RED ROCKS' AND 'PIKE'S PEAK PURPLE'

Erika Lyons, Margaret Aiken, Mara Grossman, Holly Scoggins and Joyce Latimer

Department of Horticulture, Virginia Tech, Blacksburg, VA 24061 U.S.A.

ABSTRACT

In order to develop production guidelines for this vigorous crop, we evaluated foliar and/or liner applications of plant growth retardants to two *Penstemon x mexicali* cultivars, Pike's Peak Purple and Red Rocks. Unrooted cuttings were dipped in 1,000 $mg{\cdot}L^{-1}$ IBA rooting hormone solution, stuck in 35.4 ml cells filled with a peat-based soilless substrate and allowed to root under mist in a glass greenhouse for 2 weeks. For 'Pike's Peak Purple,' liner soaks or drenches (10 ml/liner) of 16 $mg{\cdot}L^{-1}$ paclobutrazol were applied one day prior to planting. All rooted liners were planted into 1.13-L plastic pots containing a peat based soilless substrate. Foliar applications of 80 or 160 $mg{\cdot}L^{-1}$ paclobutrazol or 30 or 45 $mg{\cdot}L^{-1}$ uniconazole were made one week after planting (WAP), while 5,000 $mg{\cdot}L^{-1}$ daminozide was applied twice, at 1 and 3 WAP. Liner soak or drench caused significant stunting of plants. A subsequent study with 'Pike's Peak Purple' and 'Red Rocks' evaluated liner drenches (10 ml/liner) at 0, 4, 8, or 12 $mg{\cdot}L^{-1}$ paclobutrazol. Height and width of 'Pike's Peak Purple' plants showed a quadratic reduction in response to increasing drench rates. 'Red Rocks' was more responsive to paclobutrazol with a greater quadratic reduction in plant height or width with the 4 $mg{\cdot}L^{-1}$ drench. Although growth regulation was achieved, unacceptable finished plant form requires additional management through manual pinching or the use of chemical branching agents.

INTRODUCTION

Greenhouse and nursery growers are faced with challenges in managing the growth of herbaceous perennials in containers. Height control is a central issue because the quality of the product is often reduced when plants are overgrown in their container. So, it is crucial that growers find the proper balance of plant size and container size in order to maintain quality of the product for consumers (Kessler and Keever, 2008). Plant growth regulators (PGRs) provide an effective means to improve quality of container crops by altering their growth habit and promoting compactness.

Foliar spray concentrations of paclobutrazol or uniconazole ranging from 10 to 160 $mg{\cdot}L^{-1}$ were effective at reducing the size of four different bedding plant species, especially as the concentrations increased (Barrett and Nell, 1992). Foliar spray applications of 120 $mg{\cdot}L^{-1}$ paclobutrazol or 45 $mg{\cdot}L^{-1}$ uniconazole suppressed height in *Tradescantia virginiana* L. cultivars (White et al., 2005). Gibson and Whipker (2001) studied ornamental cabbage and kale cultivars and found that spraying 2,500 $mg{\cdot}L^{-1}$ daminozide twice produced better growth control than a single application and resulted in plants that were 21% shorter than untreated plants.

Plant growth regulators may also be applied as substrate drenches to final containers or as pre-transplant as liner soaks for effective growth control of floriculture crops. Schnelle and Barrett (2010) found that applying paclobutrazol as a soak to three different bedding plant liners resulted in decreased size when compared to the control. Paclobutrazol applied as a drench to Rhododendron (*Rhododendron catawbiense* Michx.) was more effective at inhibiting growth when applied in the liner stage before transplant rather than after transplant (Gent, 2004).

The objective of this study is to evaluate the effects of various plant growth regulators applied via various methods on reducing growth in two *Penstemon x mexicali* cultivars, Pike's Peak Purple and Red Rocks.

MATERIALS AND METHODS

For all experiments, commercially produced, unrooted *Penstemon* cuttings were dipped in 1,000 $mg{\cdot}L^{-1}$ IBA rooting hormone solution, stuck in 35.4 ml cells (size 72) filled with a peat-based soilless substrate and allowed to root under mist in a glass greenhouse for 2 weeks (Grossman et al., 2013). Liners were not pinched. Rooted liners were planted into quart plastic pots (1.13-L) or trade gallon pots (2.8-L) containing a peat based soilless substrate and received 100 to 150 $mg{\cdot}L^{-1}$ N using a commercial complete fertilizer via constant liquid feed (CLF) for the remainder of their production. Potted plants were grown in a double poly greenhouse with average daily temperatures of 19.2 to 19.7°C with a daily light integral (DLI) of 7.8 to 10.5 $mol{\cdot}m^{-2}{\cdot}d^{-1}$ depending on the experiment. Natural lighting was supplemented with HPS lights for 16 hour to ensure long day conditions.

Plants were measured for plant height (from rim of the pot to the top of the plant) and plant width (average of the largest width and the width perpendicular to the largest width). Vegetative and flower stalk heights and percent of plants flowering were measured at the final data collection. Days to open flower were counted from the day of transplant to the day of the first open flower. Plants were arranged in a completely randomized design with six to eight single plant replications per treatment. Depending on the study, data were analyzed using regression analysis or analysis of variance ($P \leq 0.05$, Student's *t*-test) and least significant difference means separation through JMP Pro 10 (SAS Institute Inc, Cary, NC).

PGR Screening Test on Penstemon 'Pike's Peak Purple'

Three different PGRs were tested. PGR applications to 'Pike's Peak Purple' included an untreated control; foliar spray applications of 80 or 160 $mg{\cdot}L^{-1}$ paclobutrazol (Piccolo 10 XC, 4.0%; Fine Americas, Inc., Walnut Creek, CA); 16 $mg{\cdot}L^{-1}$ liner drench (10 ml/liner applied one day prior to planting) or liner soak (2 min soak applied one day prior to planting) applications of paclobutrazol; 30 or 45 $mg{\cdot}L^{-1}$ foliar spray applications of uniconazole (Concise, 0.055%; Fine Americas, Inc.); or two foliar applications of 5000 $mg{\cdot}L^{-1}$ daminozide (Dazide, 85%; Fine Americas, Inc.). The second daminozide treatment was applied about two weeks after the first treatment. Foliar sprays were applied at a volume of 210 $ml{\cdot}m^{2}$ at one week after transplanting to quart pots. Plant growth was measured at 1, 3, 5, and 7 weeks after potting (WAP).

Paclobutrazol drench applications on Penstemon 'Red Rocks' and 'Pike's Peak Purple'

Four rates of paclobutrazol drenches were applied during the liner stage plants of both *Penstemon* cultivars by pouring 10 mL of the PGR solution over the media in each cell one day before the rooted liners were transplanted into trade gallon pots. The PGR drenches included an untreated control or paclobutrazol drenches at 4, 8, or 12 $mg{\cdot}L^{-1}$. Plant growth was measured at 0, 2, 4, 6, and 8 WAP.

RESULTS

PGR Screening on Penstemon 'Pike's Peak Purple'

At one week after potting (WAP), both the paclobutrazol liner soak and liner drench applications significantly reduced plant height (29% and 36%, respectively) as compared to plants in the control and not yet treated groups (Table 1). At 3 WAP, the plants subjected to any of the foliar PGR applications, except daminozide, were shorter when compared to the control with the high rates of paclobutrazol or uniconazole causing the greatest growth reductions. The plants subjected to the paclobutrazol liner soak or liner drench were stunted, 56% or 63% shorter, respectively, than control plants. Average plant width was not significantly reduced by the foliar PGR applications as compared to the control, but was reduced about 20% by the liner treatments. At 5 WAP, plants subjected to any of the PGR applications were shorter than controls but width was reduced only by the liner applications (Figure 1).

Table 1. Plant height and average width of Penstemon x mexicali 'Pike's Peak Purple' at 1 week, 3 weeks, and 5 weeks after potting. Paclobutrazol liner soaks (2 min) and drenches (10 ml/cell) were applied 1 day prior to potting, and foliar applications of the PGRs paclobutrazol, uniconazole, and daminozide were applied one week after potting.

	1 week	3 weeks		5 weeks	
PGR (mg·L^{-1})	Plant height (cm)	Plant height (cm)	Plant width (cm)	Plant height (cm)	Plant width (cm)
Control	10.9ab[z]	19.9a	12.1ab	29.8a	12.6b
80 Paclobutrazol spray	10.9ab	15.8bc	11.4ab	22.1bc	12.4b
160 Paclobutrazol spray	9.8b	12.6d	11.3b	16.1d	12.0b
30 Uniconazole spray	11.9a	16.6bc	12.1ab	23.9bc	12.5b
45 Uniconazole spray	11.1ab	14.6cd	11.3ab	20.9c	12.1b
5000 Daminozide spray	10.3ab	17.8ab	12.2a	25.6b	14.1a
16 Paclobutrazol liner drench	7.0c	7.4e	9.5c	9.6e	10.0c
16 Paclobutrazol liner soak	7.7c	8.7e	9.7c	11.3e	10.6c
P-value	<0.0001	<0.0001	<0.0001	<0.0001	<0.0001
LSD	1.7	2.4	0.9	3.7	1.2

[z]Means within a column followed by the same letter are not significantly different (Student's t-test, P<0.05, n=8).

Figure 1. *Penstemon x mexicali 'Pike's Peak Purple' plants at 5 weeks after potting (left to right): untreated control, 16 mg·L⁻¹ paclobutrazol liner soak (PAC soak), 16 mg·L⁻¹ paclobutrazol liner drench (PAC drench), 80 or 160 mg·L⁻¹ paclobutrazol foliar spray (PAC spray), 30 or 45 mg·L⁻¹ uniconazole foliar spray (UNI spray), and 5000 mg·L⁻¹ daminozide foliar spray (DAM spray) applied twice.*

At the final data collection (7 WAP), vegetative height reductions of 'Pike's Peak Purple' plants persisted with both rates of foliar applications of paclobutrazol or uniconazole as compared to the control (Figure 2, Table 2). Plants subjected to the liner soak or liner drench still exhibited stunting with 44% or 54%, respectively, reductions in plant height compared to the control plants. Flower stalk height also was reduced by all the foliar applications except daminozide as compared to the control plants. Average plant width was reduced by all PGR applications. The PGR applications had no significant effect on the number of days required to flower. However, we observed low flowering percentages in the paclobutrazol applications that caused the greatest growth reductions. Daminozide also reduced the percent of plants flowering. Beyond the stunting effects of the PGRs, the plants were not considered marketable due to poor growth habit and appearance.

Figure 2. Penstemon x mexicali 'Pike's Peak Purple' plants at 7 weeks after potting (left to right): untreated control, 16 mg·L⁻¹ paclobutrazol liner soak (PAC soak), 16 mg·L⁻¹ paclobutrazol liner drench (PAC drench), 80 or 160 mg·L⁻¹ paclobutrazol foliar spray (PAC spray), 30 or 45 mg·L⁻¹ uniconazole foliar spray (UNI spray), and 5000 mg·L⁻¹ daminozide foliar spray (DAM spray) applied twice.

Table 2. Vegetative height, flower stalk height, average width, days to flower, and percent of plants flowering of Penstemon x mexicali 'Pike's Peak Purple' at 6 weeks after foliar application of the PGRs paclobutrazol, uniconazole, or daminozide and 7 weeks after potting and application of paclobutrazol liner soak and drenches.

PGR (mg·L^{-1})	Vegetative height (cm)	Flower stalk height (cm)	Plant width (cm)	Days to flower[x]	Percent (%) plants flowering
Control	23.7a[z]	42.4a	16.8a	38	63%
80 Paclobutrazol spray	20.0b	34.0c	13.1cd	38	63%
160 Paclobutrazol spray	17.4c	27.0d	10.7ef	45	25%
30 Uniconazole spray	18.9bc	36.9bc	13.7bc	41	100%
45 Uniconazole spray	18.8bc	33.8c	12.2de	42	88%
5000 Daminozide spray	24.2a	38.4ab	14.9b	35	38%
16 Paclobutrazol liner drench	10.8e	16.3e	7.8g	47	25%
16 Paclobutrazol liner soak	13.3d	18.5e	10.3f	--	0%
P-value	<0.0001	<0.0001	<0.0001	0.3147	<0.0001
LSD	2.5	--[y]	1.5	--	--

[z]Means within a column followed by the same letter are not significantly different (Student's t-test, P<0.05, n=8).
[y]LSD not available for flowering data due to not all plants flowering in the PGR groups.
[x]Days to flower were measured from the day of transplant to the day of first open flower.

Paclobutrazol drench applications on Penstemon 'Red Rocks'

As early as 2 WAP, both plant height and width of 'Red Rocks' showed a significant linear and quadratic reduction in response to increasing paclobutrazol rate (data not presented). This correlation persisted at 4 WAP, with a quadratic response having the best fit wherein height was decreased as paclobutrazol rate increased until the PGR appeared to reach saturation point at the 8 mg·L^{-1} rate with little additional reduction in height with increasing rates (Figure 3). Average plant width was significantly reduced in a quadratic response to all of the paclobutrazol drench rates with the response saturated at 4 mg·L-1 (control width = 14.5 cm; data not presented).

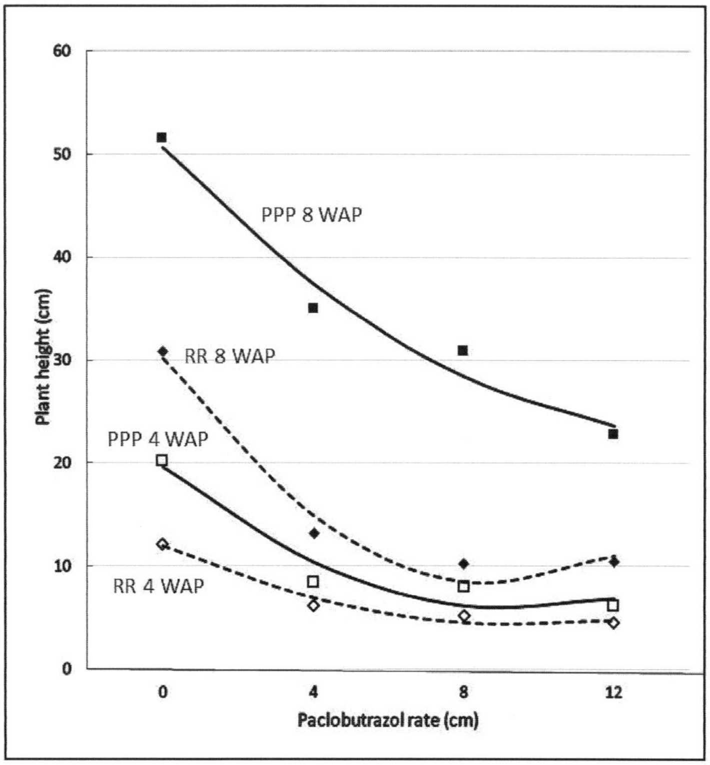

Figure 3. Plant height of Penstemon x mexicali 'Red Rocks' (RR) and 'Pike's Peak Purple' (PPP) at 4 and 8 weeks after potting. Paclobutrazol liner drenches were applied at 10 mL per plant the day prior to potting.
Regression equations:
*Red Rocks (RR) 4 WAP Ht = 11.97 – 1.596*x + 0.08281*x^2 R^2 = 0.913*
*Red Rocks (RR) 8 WAP Ht = 30.22 – 4.933*x + 0.2781*x^2 R^2 = 0.816*
*Pike's Peak Purple (PPP) 4 WAP Ht = 19.58- 2.919*x +0.1547*x^2 R^2 = 0.899*
*Pike's Peak Purple (PPP) 8 WAP Ht = 50.7 – 3.851*x + 0.1328*x^2 R^2 = 0.936*

By the final data collection at 8 WAP, height of 'Red Rocks' plants was reduced 57% by the 4 mg·L-1 paclobutrazol rate, 67% by the 8 mg·L-1 rate, and 66% by the 12 mg·L-1 rate when compared to the control (Figure 4). Average width also had a quadratic response to increasing paclobutrazol drench rates: with 55%, 70% or 70% reductions at 4, 8 or 12 mg·L-1 paclobutrazol, respectively (control width = 29.4 cm; data not presented). All plants were stunted by all rates and did not recover from the applications. However, all plants flowered with only 3 to 5 day delays relative to control plants (data not presented).

Figure 4. Penstemon x mexicali 'Red Rocks' plants at 8 weeks after potting and application of paclobutrazol liner drenches at (left to right) 0, 4, 8 or 12 mg·L^{-1} applied at 10 ml per plant in the liner tray one day prior to potting.

Paclobutrazol drench applications on Penstemon 'Pike's Peak Purple'

As with 'Red Rocks' growth reductions were significant at 2 WAP with a quadratic response to increasing rates (data not presented). At 4 WAP, reductions in plant height were saturated at the 4 mg·L^{-1} paclobutrazol drench rate with a 58% reduction relative to control plants (Figure 3). Reductions in average width also showed a quadratic response to increasing rate but was less severely affected than height, e.g., width was reduced 37% by either the 4 or 8 mg·L^{-1} drenches and 47% by the 12 mg·L^{-1} drench (control width = 17.4 cm; data not presented).

At the final data collection at 8 WAP, the 'Pike's Peak Purple' plants were exhibiting some recovery from the early stunting with height reductions of 32%, 40%, or 56% with the 4, 8, or 12 mg·L^{-1} drenches, respectively (Figure 3). However, average width reductions were slightly greater than at 4 WAP with 46%, 56% or 66% with the 4, 8, or 12 mg·L^{-1} drenches, respectively, as compared to control plants (control width = 34.6 cm; data not presented). As with 'Red Rocks,' all plants flowered with only the highest paclobutrazol rate causing a significant, 5 day, delay.

Figure 5. Penstemon x mexicali 'Pike's Peak Purple' plants at 8 weeks after potting and application of paclobutrazol liner drenches at (left to right) 0, 4, 8 or 12 mg·L^{-1} applied at 10 ml per plant in the liner tray one day prior to potting.

DISCUSSION

It is clear from the screening trial that the paclobutrazol liner drench or liner soak were the most effective treatments at growth reduction. Blanchard and Runkle (2007) reported that applying liner dips/soaks of paclobutrazol at 4, 8, or 16 mg·L^{-1} to different bedding plants resulted in reduced stem elongation and shorter plants. However, in our study the *Penstemon* plants subjected to these application rates were over-regulated and not marketable due to severe stunting. In contrast, the foliar spray applications of paclobutrazol, uniconazole, and daminozide did not visibly reduce height and width enough for desirable growth control. However, Starman (1990) found that foliar applications of uniconazole and daminozide were effective at reducing height in chrysanthemums. Higher concentrations or multiple applications of these sprays may be needed in order to provide a sufficient growth regulation to create a marketable plant.

All three concentrations (4, 8 or 12 mg·L^{-1}) of the paclobutrazol drench reduced height and average width in both of the *Penstemon* cultivars. Over-regulation and stunting were evident in these treatment groups as well, especially with the higher paclobutrazol concentrations. In a study by Ruter (1996), different paclobutrazol application methods were effective at reducing growth in 'New Gold' lantana (*Lantana camara* L.), but lower rates were recommended as a result of too much growth control. The results of our study suggest that drench rates should be reduced further in order to prevent over-regulation and stunting of *Penstemon*. Whipker and Hammer (1997) found that 2 to 4 mg or 4 to 8 mg of a paclobutrazol drench application produced marketable dahlia (*Dahlia variabilis* Willd.) plants depending on the cultivar.

The paclobutrazol drench also produced darker green and glossy foliage in 'Red Rocks' but not in 'Pike's Peak Purple.' It is clearly evident that liner drenches of paclobutrazol are effective at reducing growth in *Penstemon* cultivars. However, it is critical that preliminary tests be performed before application of this PGR on *Penstemon* in order to find the ideal concentration that reduces growth but does not cause over-regulation to the crop. The use of a pinching treatment and/or the application of chemical branching agents may also improve plant growth habit by filling out the plant.

LITERATURE CITED

Barret, J.E. and T.A. Nell. 1992. Efficacy of paclobutrazol and uniconazole on four bedding plant species. HortScience 27(8):896-897.

Blanchard, M.G. and E.S. Runkle. 2007. Dipping bedding plant liners in paclobutrazol or uniconazole inhibits subsequent stem extension. HortTechnology 17(2): 178-182.

Gent, M.P.N. 2004. Efficacy and persistence of paclobutrazol applied to rooted cuttings of rhododendron before transplant. HortScience 39(1):105-109.

Gibson, J.L. and B.E. Whipker. 2001. Ornamental cabbage and kale growth responses to daminozide, paclobutrazol, and uniconazole. HortTechnology 11(2):226-230.

Grossman, M., J. Freeborn, H. Scoggins, and J. Latimer. 2013. Timing of benzyladenine and sodium dikegulac application affects branching of liners and finished herbaceous perennial plants. HortScience 48:875-882.

Kessler, Jr, J.R. and G.J. Keever. 2008. Plant growth retardants affect growth and flowering of *Achillea* x 'Coronation Gold.' J. Environ. Hort. 26(1):24-28.

Ruter, J.M. 1996. Paclobutrazol application method influences growth and flowering of 'New Gold' Lantana. HortTechnology 6(1): 19-20.

Schnelle, R.A. and J.E. Barrett. 2010. Paclobutrazol concentration and substrate moisture status impact efficacy of liner dips for size control of three bedding plants. HortTechnology 20(4):735-739.

Starman, T.W. 1990. Whole-plant response of chrysanthemum to uniconazole foliar sprays or medium drenches. HortScience 25(8): 935-937.

Whipker, B.E. and P.A. Hammer. 1997. Efficacy of ancymidol, paclobutrazol, and uniconazole on growth of tuberous-rooted dahilias. HortTechnology 7(3): 269-273.

White, S.A., H.L. Scoggins, R.P. Marini, and J.G. Latimer. 2005. Multivariate repeated measures analysis of plant growth regulators on *Tradescantia virginiana*. HortScience 40(2):404-408.

BRASSINOSTEROIDS PROMOTE GERMINATION AND SEEDLING GROWTH OF RICE

Yi Su[1,2], Liangliang Chen[1], Jianhua Tong[1], Huiqun Wang[1], Wanhuang Lin[1], Langtao Xiao[1,2]

[1]Hunan Provincial Key Laboratory of Phytohormones and Growth Development, Hunan Agricultural University, Changsha, China

[2]Hunan Co-Innovation Center for Utilization of Botanical Functional Ingredients, Changsha, China

Abstract

Brassinosteroids (BRs), a phytohormone category of more than 50 naturally occurring compounds with diverse physiological roles, have been found to regulate multiple processes of cell division, cell elongation, vascular bundle differentiation, root development, flowering, senescence, photosynthesis, and stress tolerance in plants. Therefore, study on BR application in staple crops is of great significance in agriculture. In this research, the effects of BR on rice seed germination and seedling growth have been examined. Firstly, rice seeds were soaked in 10 ppm BR for 24 hours, and then after the seeds were transferred to a growth chamber in darkness at 30 °C for 2 days. The results showed that the treated rice seeds were able to germinate more coincidently comparing to the control. Similarly, when the seeds were pretreated with ABA to suppress germination, treatment with 10 ppm BR was able to counteract the ABA effect by promoting seeds germination. It was also found that the germination rate, the bud and radical growth of ABA treated seeds have been reduced for about 50%, while the BR treatment of the ABA pretreated seeds obtained normal germination and bud/radical growth comparing to the control. Furthermore, BR was able to increase root activity, photosynthesis efficiency and activity of several antioxidant enzymes. Since BR can promote seed germination and seedling growth of rice, soaking rice seeds with BR would be an ideal method for rice seed pretreatment in the production practice.

INTRODUCTION

For plants, the appropriate timing of growth occurs in response to specific environmental signals as regulated mainly by various phytohormones (Jaillais and Chory, 2010; Finet and Jaillais, 2012). Brassinosteroids (BRs), a phytohormone category of more than 70 naturally occurring compounds with diverse physiological roles, have been found to regulate multiple processes of cell division, cell elongation, vascular bundle differentiation, root development, flowering, senescence, photosynthesis, and responses to the environment in plants (Heyman et al., 2013; Clouse, 2014; Belkhadir and Jaillais, 2015). Defects in BR biosynthesis or signaling result in physiological and phenotypic abnormalities, such as reduced seed germination, extreme dwarfism, photomorphogenesis in the dark, altered distribution of stomata, delayed flowering and male sterility. Therefore, study on BR application in staple crops is of great significance in agriculture (Vilarrasa-Blasi et al., 2014; Wang et al., 2014; Zhu et al., 2013).

Fortunately, the molecular mechanisms of BR signal transduction and BR mediated regulation of gene expression and plant development have been dissected in detail by studies based on molecular genetics, biochemistry, proteomics and genomics, making the BR signaling pathway one of the best understood signal transduction pathways in plants, providing valuable action mechanism information for the application research of BR in crops. Various BR mutants have been identified by profound morphological phenotypes such as dwarfism, delayed flowering, reduced male fertility, altered vascular development and impaired photomorphogenesis (Clouse, 1996; Belkhadir and Chory, 2006; Jaillais and Vert, 2012; Fridman and Savaldi-Goldstein, 2013). Although the BR signaling pathway is known mostly for its primary actions on cell expansion, it may confer increased adaptation to various biotic and abiotic stresses (Belkhadir and Chory, 2006). The regulation of these processes by BR might contribute to plant adaptation to a certain environment (Fridman and Savaldi-Goldstein, 2013). BR signaling begins at the cell surface, where BR is perceived by a small family of plasma membrane-localized leucine-rich repeat receptor kinases (LRR-RKs). BRASSINOSTEROID INSENSITIVE 1 (BRI1) is the first characterized member of this family of receptors and account for most of the BR binding activity in *Arabidopsis* (Cano-Delgado et al., 2004). BR perception at the cell surface drives a fully connected signaling cascade that terminates in the nucleus (Belkhadir and Chory, 2006). Downstream of the receptor, BR signals are modulated by key master regulators of the pathway. Firstly, a phosphorylation-dephosphorylation cascade involving the GSK3-like kinase BRASSINOSTEROID INSENSITIVE 2 (BIN2) controls the transduction of BR signals to the downstream plant-specific transcription factors BRASSINAZOLE-RESISTANT1 (BRZ1) and BR-INSENSITIVE-EMS-SUPPRESSOR1 (BES1)/BZR2 (Wang et al., 2014). Secondly, BIN2 also transduces BR signals on a large repertoire of physiological targets that act in various aspects of plant development (Guo et al., 2013; Wang et al., 2014). Thirdly, BZR1/BES1 integrate, in successive waves, BR signals on the promoter of several thousands of genes involved in other signaling pathways (Guo et al., 2013). The BR transcriptional network, through the action of BES1 and BZR1, provides points of crosstalk for different pathways, such as cell elongation, flowering, senescence, morphogenesis and seed germination (Zhu et al., 2013). In recent years, significant progress has been made in understanding how BES1 and BZR1 can interact with other transcription factors to modulate the growth processes controlled by light, GA and Auxin (Fridman and Savaldi-Goldstein, 2013; Wang et al., 2014). Current models propose that GA and BR influence cell elongation through tangled but distinct signaling pathways (Wang et al., 2014). In this research, the effects of BR on rice seed germination and seedling growth have been examined, and this will support the application in rice cultivation. Moreover, rice seeds were treated with BR and ABA simultaneously for determining the interaction between BR and ABA at the stage of seed germination.

MATERIALS AND METHODS

For germination tests, rice seeds were soaked in different solutions (10 ppm BR, water, 50 mg/L ABA and 50 mg/L ABA plus 10 ppm BR respectively) for 24 hours. Then the seeds were transferred to a growth chamber in darkness at 30 ^{o}C for 2 days. The germination rate was counted in the first day and the root length was measured in the second day. For plant growth analysis, germinated rice seeds were sowed in vermiculites irrigated with Hoagland solution and cultured in growth chamber under 16 hours light and 8 hours darkness at 30 ^{o}C. The rice seedlings were treated with 10 ppm BR and water respectively by spraying in the seventh day. The height and dry weight of seedlings were measured two weeks later. Moreover, root activities were measured through TTC method and activities of several antioxidant enzymes were detected by using commercial kits.

RESULTS

BR promotes rice seed germination

Rice seeds were directly soaked with BR solution, water, ABA solution and ABA plus BR solution respectively and then the seed germination was analyzed. The results showed that BR could promote seed germination and the BR treated rice seeds were able to germinate more coincidently when comparing to the control (Figure 1A-E). Similarly, when the seeds were pretreated with ABA to suppress germination, treatment with 10 ppm BR was able to counteract the ABA effect by promoting seeds germination (Figure 2A and B). It was found that the germination rate, the bud and root growth of ABA treated seeds have been reduced for about 50%, while the BR treatment of the ABA pretreated seeds obtained normal germination and unique bud/root growth when comparing to the control (Figure 2C).

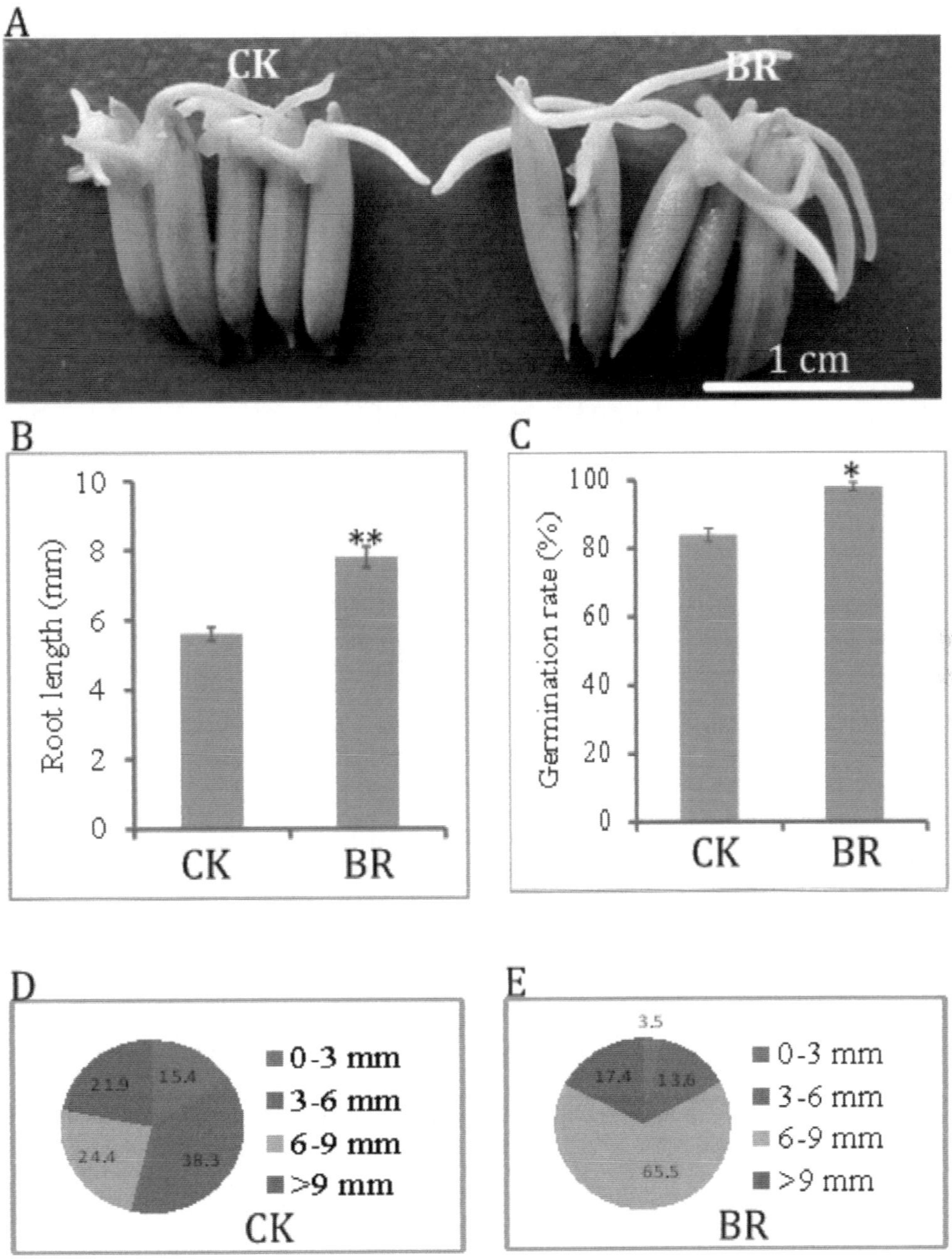

Figure 1 Effects of BR on rice seed germination. (A) germinated seeds, (B) root length and (C) germination rate in the 2nd day after germination, and percentages of different root lengths when treating without BR (D) and with BR (E).

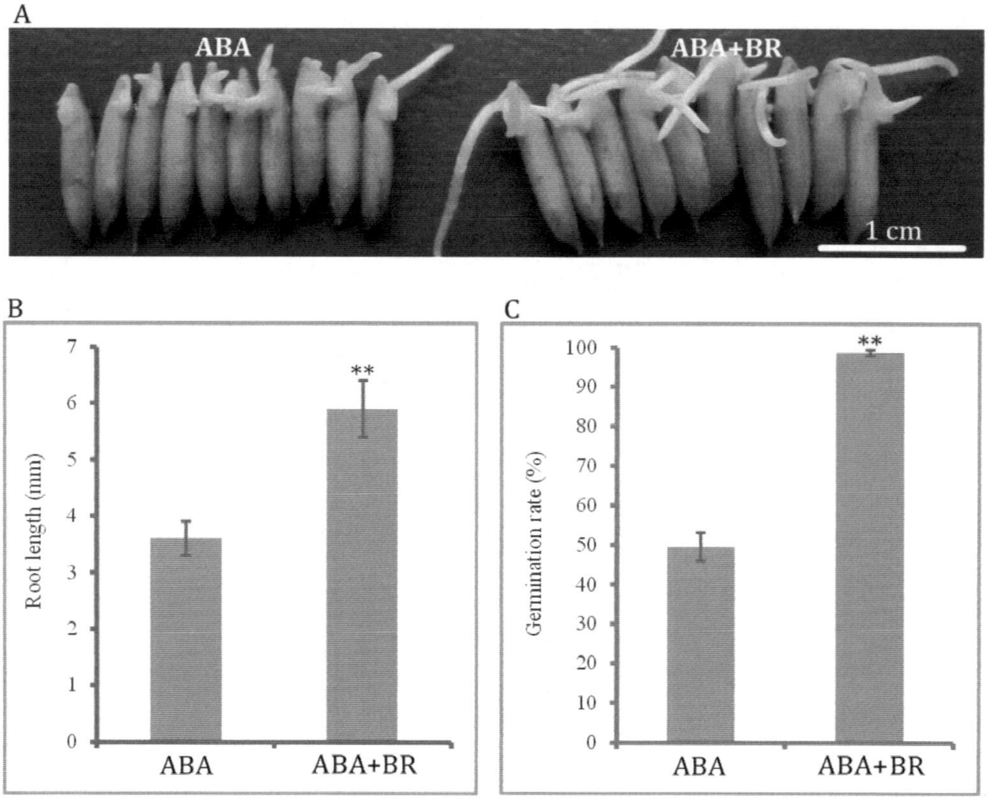

Figure 2 Pretreated rice seeds with brassinosteroid plus 50 mg/L ABA existence. (A) the second day after germination, (B) root length and (C) germination rate at the second day after germination.

BR PROMOTES RICE SEEDLINGS GROWTH

The germinated rice seeds were transplanted in vermiculites. Two weeks later, the growth of rice seedlings were analyzed. The results showed that BR was able to significantly promote rice seedling growth (Figure 3A-C). The heights and dry weights of seedlings from BR soaked seeds were significantly increased when comparing to the control (Figure 3A-C). As shown in figure 4, the increased root activity was one of the reasons why BR is able to promote rice growth.

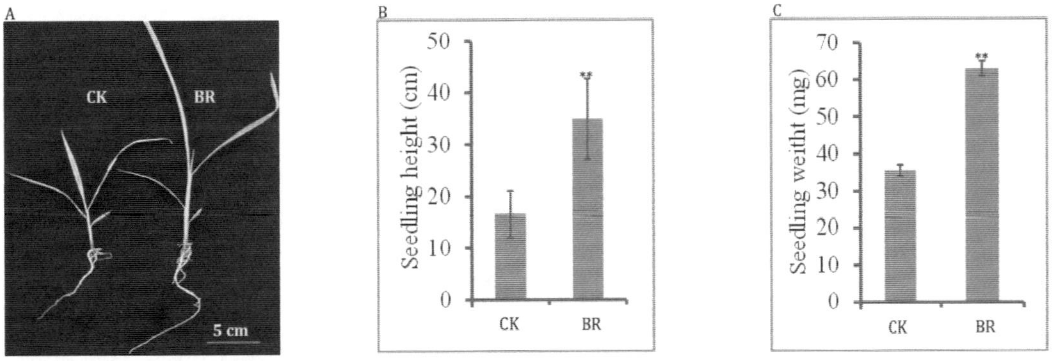

Figure 3 Treated rice seedlings with brassinosteroid. (A) rice seedlings after treatment, (B) height and (C) dry weight of seedlings after treatment.

Brassinosteroid affects activities of antioxidant enzymes

One of BR's roles is the response to environmental stresses. Thus activities of several typical antioxidant enzymes were detected in rice leaves. The results showed that BR was able to increase activities of POD, CAT and SOD significantly (Figure 5).

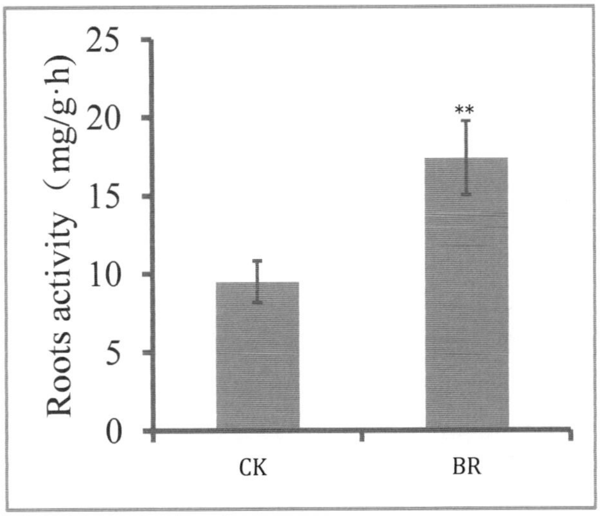

Figure 4 Root activities of treated rice seedlings.

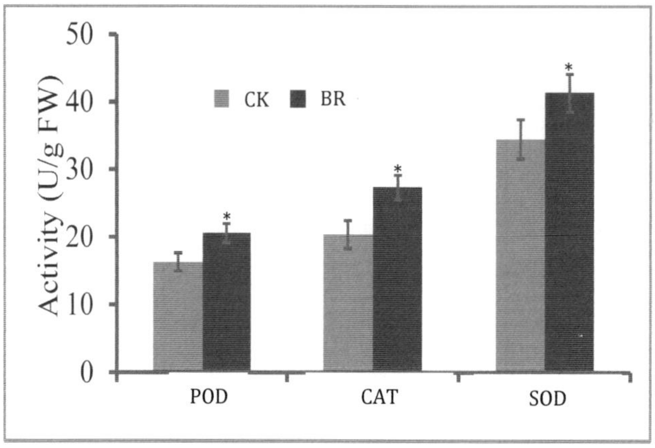

Figure 5 Antioxidant enzymes activities of rice seedlings.

DISCUSSION

BR is able to cross-talk with other signaling pathways. Nearly all BR-regulated processes are also regulated by other hormonal or environmental signals (Depuydt and Hardtke, 2011). For example, suppression of photomorphogenesis in the dark requires BR and two other hormones: GA and auxin (Li et al., 1996; Szekeres et al., 1996; Lau and Deng, 2010). The bulk of evidence shows that BR antagonises light signals, has similar physiological effects to GA, and interacts synergistically with auxin during cell elongation and gene expression (Depuydt and Hardtke, 2011; Lau and Deng, 2010). Recent studies have uncovered the molecular mechanisms underlying the interactions between BR and other phyohormones (Bai et al., 2012; Gallego-Bartolomé et al., 2012; Li et al., 2012; Luo et al., 2010; Oh et al., 2012). BR, auxin and GA, belonging to growth-promoting phytohormones, have similar effects on various developmental processes throughout the life cycle of plants (Depuydt and Hardtke, 2011). GA binds to its receptor GIBBERELLIN INSENSITIVE DWARF1 (GID1) to induce ubiquitylation and degradation of DELLA family proteins, which bind to several transcription factors, including PIFs, to inhibit their DNA binding. GA induces degradation of DELLA to derepress PIFs (de Lucas et al., 2008; Feng et al., 2008). Until recently, the effects of BR and GA have been considered additive (Depuydt and Hardtke, 2011), and their signaling pathways have been proposed to act on largely non-overlapping transcriptional responses (Nemhauser et al., 2006). However, recent studies demonstrated an interdependent relationship and a direct interaction between the BR and GA signaling pathways (Bai et al., 2012; Gallego-Bartolomé et al., 2012; Li et al., 2012). Microarray analyses showed that BR and auxin also activate overlapping sets of genes, while genetic and physiological studies showed synergistic and interdependent interactions between the two hormones in a wide range of developmental contexts, such as hypocotyl elongation and vascular bundle patterning (Depuydt and Hardtke, 2011; Ibañes et al., 2009; Nemhauser et al., 2004). Several observations suggest potential molecular mechanisms of mediating BR-auxin interactions. Firstly, BZR1 regulates many genes involved in auxin synthesis, transport and signaling (Sun et al., 2010) and auxin activates BR biosynthetic genes and increases BR levels (Chung et al., 2011; Mouchel et al., 2006; Yoshimitsu et al., 2011). Secondly, BIN2 phosphorylates an auxin-response factor (Vert et al., 2008) while BZR2 and

ARF5 bind to the same BR/auxin-response promoter (Walcher and Nemhauser, 2012). Thirdly, BR and auxin responses are integrated through the actin cytoskeleton, which is not only regulated by both BR and auxin but also feedback regulates auxin transport and BR signaling (Lanza et al., 2012). The contributions of these mechanisms to the physiology of BR-auxin interaction remain to be evaluated further. Interestingly, in this research we found that a certain concentration of BR was able to relieve the ABA repression although ABA alone seriously represses the rice seed germination. This result indicated that BR may interact with ABA in jointly regulating the rice seed germination while the molecular mechanism deserves further exploration. Overall, BR shows high bioactivity in promoting seed germination and seedling growth of rice, thus soaking rice seeds with BR will be an ideal method for rice seed pretreatment in the production practice.

ACKNOWLEDGMENT

This work was supported by Natural Science Foundation of China, grants 91317312 and 31570372.

REFERENCES

Bai, M. Y., Shang, J. X., Oh, E., Fan, M., Bai, Y., Zentella, R., Sun, T. P. and Wang, Z. Y. (2012). Brassinosteroid, gibberellin and phytochrome impinge on a common transcription module in *Arabidopsis*. Nat. Cell Biol. 14, 810-817.

Belkhadir, Y., and Chory, J.(2006). Brassinosteroid signaling: a paradigm for steroid hormone signaling from the cell surface. Science 314, 1410-1411.

Belkhadir, Y., and Jaillais, Y. (2015). The molecular circuitry of brassinosteroid signaling. New Phytol. 206, 522-540.

Cano-Delgado, A., Yin, Y. H., Yu, C., Vafeados, D., Mora-Garcia, S., Cheng, J. C., Nam, K. H., Li, J., and Chory, J. (2004). BRL1 and BRL3 are novel brassinosteroid receptors that function in vascular differentiation in *Arabidopsis*. Development 131, 5341-5351.

Chung, Y., Maharjan, P. M., Lee, O., Fujioka, S., Jang, S., Kim, B., Takatsuto, S., Tsujimoto, M., Kim, H., Cho, S., Park T., Cho, H., Hwang, I. and Choe, S. (2011). Auxin stimulates DWARF4 expression and brassinosteroid biosynthesis in *Arabidopsis*. Plant J. 66, 564-578.

Clouse, S. Brassinosteroid signaling in plants. (2014). Plant Sciences, 2: 291-312.

Clouse, S. D., Langford, M. and McMorris, T. C. (1996). A brassinosteroid insensitive mutant in *Arabidopsis thaliana* exhibits multiple defects in growth and development. Plant Physiol. 111, 671-678.

de Lucas, M., Davière, J. M., Rodríguez-Falcón, M., Pontin, M., Iglesias-Pedraz, J. M., Lorrain, S., Fankhauser, C., Blázquez, M. A., Titarenko, E. and Prat, S. (2008). A molecular framework for light and gibberellin control of cell elongation. Nature 451, 480-484.

Depuydt, S. and Hardtke, C. S. (2011). Hormone signalling crosstalk in plant growth regulation. Curr. Biol. 21, R365-R373.

Feng, S. H., Martinez, C., Gusmaroli, G., Wang, Y., Zhou, J. L., Wang, F., Chen, L. Y., Yu, L., Iglesias-Pedraz, J. M., Kircher, S., Schäfer, E., Fu, X. D., Fan, L. M. and Deng, X. W. (2008). Coordinated regulation of *Arabidopsis thaliana* development by light and gibberellins. Nature 451, 475-479.

Finet, C. and Jaillais, Y. (2012). Auxology: when auxin meets plant evo-devo. Dev. Biol. 369, 19-31.

Fridman, Y. and Savaldi-Goldstein, S. (2013). Brassinosteroids in growth control: how, when and where. Plant Sci. 209, 24-31.

Gallego-Bartolomé, J., Minguet, E. G., Grau-Enguix, F., Abbas, M., Locascio, A., Thomas, S. G., Alabadí, D. and Blázquez, M. A. (2012). Molecular mechanism for the interaction between gibberellin and brassinosteroid signaling pathways in *Arabidopsis*. Proc. Natl. Acad. Sci. USA 109, 13446-13451.

Guo, H. Q., Li, L., Aluru, M., Aluru, S. and Yin, Y. H. (2013). Mechanisms and networks for brassinosteroid regulated gene expression. Curr. Opin. Plant Biol. 16, 545-553.

Heyman, J., Cools, T., Vandenbussche, F., Heyndrickx, K, Leene, J., Vercauteren, I., Vanderauwera, S., Vandepoele, K., De Jaeger, G., Van Der Straeten, D., and De Veylder, L. (2013) ERF115 controls root quiescent center cell division and stem cell replenishment. Science 342, 860-863.

Ibañes, M., Fàbregas, N., Chory, J. and Caño-Delgado, A. I. (2009). Brassinosteroid signaling and auxin transport are required to establish the periodic pattern of *Arabidopsis* shoot vascular bundles. Proc. Natl. Acad. Sci. USA 106, 13630-13635.

Jaillais Y, and Chory J. 2010. Unraveling the paradoxes of plant hormone signaling integration. Nat. Struct. Mol. Biol. 17, 642-645.

Jaillais, Y. and Vert, G. (2012). Brassinosteroids, gibberellins and light-mediated signalling are the three-way controls of plant sprouting. Nat. Cell Biol. 14, 788-790.

Lanza, M., Garcia-Ponce, B., Castrillo, G., Catarecha, P., Sauer, M., Rodriguez-Serrano, M., Páez-García, A., Sánchez-Bermejo, E., Mohan, T. C., Leo del Puerto, Y., Sandalio, L.M., Paz-Ares, J. and Leyva, A. (2012). Role of actin cytoskeleton in brassinosteroid signaling and in its integration with the auxin response in plants. Dev. Cell 22, 1275-1285.

Lau, O. S. and Deng, X. W. (2010). Plant hormone signaling lightens up: integrators of light and hormones. Curr. Opin. Plant Biol. 13, 571-577.

Li, J., Nagpal, P., Vitart, V., McMorris, T. C. and Chory, J. (1996). A role for brassinosteroids in light-dependent development of *Arabidopsis*. Science 272, 398-401.

Li, Q. F., Wang, C., Jiang, L., Li, S., Sun, S. S. and He, J. X. (2012). An interaction between BZR1 and DELLAs mediates direct signaling crosstalk between brassinosteroids and gibberellins in *Arabidopsis*. Sci. Signal. 5, ra72.

Luo, X. M., Lin, W. H., Zhu, S. W., Zhu, J. Y., Sun, Y., Fan, X.Y. , Cheng, M. L., Hao, Y. Q., Oh, E., Tian, M. M., Liu, L. J., Zhang, M., Xie, Q., Chong, K., and Wang, Z. Y. (2010). Integration of light- and brassinosteroid signaling pathways by a GATA transcription factor in *Arabidopsis*. Dev. Cell 19, 872-883.

Mouchel, C. F., Osmont, K. S. and Hardtke, C. S. (2006). BRX mediates feedback between brassinosteroid levels and auxin signalling in root growth. Nature 443, 458-461.

Nemhauser, J. L., Hong, F. X. and Chory, J. (2006). Different plant hormones regulate similar processes through largely nonoverlapping transcriptional responses. Cell 126, 467-475.

Nemhauser, J. L., Mockler, T. C. and Chory, J. (2004). Interdependency of brassinosteroid and auxin signaling in *Arabidopsis*. PLoS Biol. 2, E258.

Oh, E., Zhu, J. Y. and Wang, Z. Y. (2012). Interaction between BZR1 and PIF4 integrates brassinosteroid and environmental responses. Nat. Cell Biol. 14, 802-809.

Sun, Y., Fan, X. Y., Cao, D. M., Tang, W., He, K., Zhu, J. Y., He, J. X., Bai, M. Y., Zhu, S., Oh, E. Patil, S., Kim, T.W., Ji, H. K., Wong, W. H., Rhee, S.Y., and Wang, Z.Y. ((2010). Integration of brassinosteroid signal transduction with the transcription network for plant growth regulation in *Arabidopsis*. Dev. Cell 19, 765-777.

Szekeres, M., Németh, K., Koncz-Kálmán, Z., Mathur, J., Kauschmann, A., Altmann, T., Rédei, G. P., Nagy, F., Schell, J. and Koncz, C. (1996). Brassinosteroids rescue the deficiency of CYP90, a cytochrome P450, controlling cell elongation and de-etiolation in *Arabidopsis*. Cell 85, 171-182.

Vert, G., Walcher, C. L., Chory, J. and Nemhauser, J. L. (2008). Integration of auxin and brassinosteroid pathways by Auxin Response Factor 2. Proc. Natl. Acad. Sci. USA 105, 9829-9834.

Vilarrasa-Blasi J., Gonzalez-Garcia M.P., Frigola D., Fabregas-Vallve, N., Alexiou, K. G., Lopez-Bigas, N., Rivas, S., Jauneau, A., Lohmann, J. U., Benfey, P.N., Ibanes, M., and Cano-Delgado, A. I. (2014) Regulation of plant stem cell quiescence by a brassinosteroid signaling module. Dev. Cell 30, 36-47.

Walcher, C. L. and Nemhauser, J. L. (2012). Bipartite promoter element required for auxin response. Plant Physiol. 158, 273-282.

Wang W, Bai M. Y. and Wang Z. Y. (2014). The brassinosteroid signaling network-aparadigm of signal integration. Curr. Opin. Plant Biol. 21c, 147-153.

Yoshimitsu, Y., Tanaka, K., Fukuda, W., Asami, T., Yoshida, S., Hayashi, K., Kamiya, Y., Jikumaru, Y., Shigeta, T., Nakamura, Y., Matsuo, T., and Okamoto, S. (2011). Transcription of DWARF4 plays a crucial role in auxin-regulated root elongation in addition to brassinosteroid homeostasis in *Arabidopsis thaliana*. PLoS ONE 6, e23851.

Zhu J.Y., Sae-Seaw J., and Wang Z.Y. (2013). Brassinosteroid signalling. Development 140, 1615-1620.

ETHYLENE RESPONSIVE FACTOR GENES PROVIDE A KEY FOR IMPROVING TOLERANCE TO CHILLING STRESS IN HARVESTED PEPPER

Mehea Park, Sunyoung Shin, Yeoungseuk Bae, Jungsoo Lee, Hyunjin Choi

Postharvest Research Team, National Institute of Horticultural and Herbal Science, Wanju-gun, Jeollabuk-do, Korea, 565-852

ABSTRACT

Susceptibility to damage during cold storage increases postharvest loss and reduces the storage life of subtropical plants including pepper (*Capsicum annuum* L.)undergo chilling injury. Interactions between ethylene and jasmonate may fine-tune plant responses to environmental stress. To investigate the ways in which ethylene and jasmonate, or interactions between their pathways, regulate chilling stress at a molecular level, we performed the global expression profiling of peppers maintained in cold storage, and gene expression changes in relation to chilling tolerance upon hormone treatment. Transcription analysis revealed that methyl jsamonate (MeJA) treatment upregulated genes for peroxidase and allene oxide synthase, compounds related to stress responses. The ethylene-responsive factor (ERF) family and MAP kinase, involved in ethylene signaling transcription factors, were also regulated by MeJA. Functional analysis revealed that *ERF1* and *ERF2* from pepper conferred cold resistance and high antioxidant capacity during cold storage. Interestingly, *ERFs* expression in harvested peppers was linearly up regulated by MeJA during long-term cold storage. Thus, MeJA enables peppers to respond to cold stress and ethylene signaling, which could help to prevent chilling injury. *ERFs* could be key elements in the integration of jasmonate and ethylene signals for regulating defense response genes in harvested pepper during cold storage. These findings provide a genetic foundation for preventing chilling injury in subtropical crops

INTRODUCTION

Plant responses to biotic and abiotic stresses such as cold, ultraviolet radiation, and drought depend on interactions between hormone signaling pathways rather than on the independent contributions of individual hormones (McDowell and Dangl, 2000). Cold storage is generally the most effective technology for maintaining the quality of postharvest horticultural crops by inhibiting respiration. Susceptibility to damage during cold storage increases postharvest loss and reduces the storage life of chilling-sensitive crops like pepper.

Peppers (*Capsicum annuum)* are important crops nutritionally and commercially due to their abundant vitamin contents and many demands of consumers. However, peppers below 7°C, the temperature of cold storage for long term storing, may be troubled with chilling injury and then result in deterioration of the calyx, seed browning and sheet pitting (Ryall and Pentzer 1982; Özden and Bayindirli 2002).

The synthetic plant growth regulator 1-methylcyclopropene (1-MCP) is thought to inhibit ethylene-dependent processes by interacting with ethylene receptors (Sisler and Serek 1997). It has been reported that 1-MCP delays ripening and extends shelf-life of numerous climacteric fruits by affecting ethylene production, color change and respiration (Ziliotto et al. 2008; Larrigaudiere et al. 2004). As well as, there are several reports demonstrated that 1-MCP was effective on reduction of chilling injury and activation of antioxidant enzymes (Cao et al. 2009; Yuan et al. 2010).

Jasmonate (JA), a plant signaling hormone, is important to developmental processes such as germination, fertility and senescence (Collén et al. 2006). Besides, the mechanisms by JA contribute to defense reactions for stress conditions like as wounding, extreme temperature and pathogen infection (Anderson et al. 2004). Several studies have reported that MeJA (methyl jasmonate) activated antioxidant metabolic pathways and defense mechanisms in various crops (Wang et al. 2009; Parra-Lobato et al. 2009; Sasaki-Sekimoto et al. 2005; Sasaki et al. 2001). It also has been reported that MeJA conferred chilling tolerance by inducing antioxidant defense mechanism and then reduced surface pitting (Fung et al. 2004). However, few repots have been examined mechanism about how MeJA conferring chilling tolerance based on gene expression.

To understanding how postharvest crops response correctly to chilling stress during cold storage, it is essential to investigate how each role of ethylene and jasmonate or interaction between their pathways is regulated chilling stress at molecular level, which is a necessary step to rationally overcome chilling injury. However, global expression profiling of pepper fruits in cold storage, effective postharvest management which is of great value commercially, as well as the expression changes upon treatment of substance according chilling tolerance was yet to report.

In present study, we confirmed that MeJA and 1-MCP conferred cold tolerance to pepper fruits and identified global expression changes of transcriptome in cold-, MeJA-, 1-MCP-treated and non-treated peppers during cold storage at 0°C. These results will facilitate to screening of genes improving cold tolerance and establishing genetic foundation to overcome chilling injury derived from cold storage.

MATERIALS AND METHODS

Plant materials and substance treatment

Peppers (*Capsicum annuum*. cv Super Bigarim) grown in greenhouse of National Institute of Horticultural and Herbal Science was use for the experiments. Freshly harvested peppers were divided four groups. Two groups were treated 50 µm MeJA and 1 ppm 1-MCP at 20°C for 24 hours, respectively, with another one group. And the other group was used instantly as 'non-treated peppers' sample. And then, above three groups moved into 0°C cold chamber, chilling injury of each group of peppers during cold storage was observed for 1 month.

RNA extraction and microarray analysis

RNA was extracted from non-treated peppers and MeJA-, 1-MCP-, cold-treated peppers stored 0°C for 24 hours using an RNeasy®Plant Mini Kit (Qiagen). And, quantified RNA was reverse transcribed in cDNA with Superscript™ III reverse transcriptase (Invitrogen) according to the manufacturer's instructions. Chip hybridizations were performed by GreenGene Bio Tech Inc. (Korea) using *Capsicum annuum* 135K microarray GeneChip (Roche Nimblegen). Gene set enrichment analysis was performed using AgriGO analysis tools (http://bioinfo.cau.edu.cn/agriGO/).

Quantitative RT-PCR

Real-time qRT-PCR was performed on a Rotor-Gene Q real-time PCR system (Qiagen) using the Rotor-gene SYBR Green RT-PCR Kit (Qiagen) according to manufacturer's instruction cDNA was synthesized as described above, and then each gene was amplified by PCR using Phire Hot StartII DNA Polymerase (Thermo Scientific). 50 ng of total cDNA was used as a template to amplify with gene specific primer sets. The PCR amplicon of initiation factor 4E was used as a housekeeping control.

Redox status

Transformed *E. coli* cells were cultured for 3 h at 37°C until reaching mid-log phase and were then exposed to 1.5 mM H_2O_2 for 1 h. Cultured cells (50 mL) were harvested, and total protein was extracted using a Qproteome Bacterial Protein Prep Kit (Qiagen) according to the manufacturer's instructions. The lysate supernatant was used immediately for detection of hydrogen peroxide content. The intracellular hydrogen peroxide level was determined by ferrous ion oxidation in the presence of a ferric ion indicator, xylenol orange. Protein extract (100 μL) was added to 1 ml of FOX reagent containing 0.1 mM xylenol orange, 0.25 mM ammonium ferrous sulfate, 100 mM sorbitol, and 25 mM sulfuric acid. The mixture was incubated at room temperature for 30 min or more, and then centrifuged to remove any flocculated material before measuring absorbance at 560 nm. Relative hydrogen peroxide levels were represented by nmol per mg protein. Protein concentrations were quantified using Protein Dye Reagent (Bio-Rad) according to Bradford.

Phenotypic analysis of transformed E. coli

E. coli BL21 (DE3) cells transformed either with pET-28a(+) (empty vector) or with cloned plasmid (pET-*Ca*NAC1, pET-*Ca*WRKY40, pET-*Ca*ERF1, and pET-*Ca*ERF2) were grown in fresh LB-medium containing kanamycin (50 μg/mL) under continuous shaking at 37°C. After the mid-log phase (A660≑0.3~0.6) was reached, the cells were streaked on LB agar plates. The plates were incubated at 7°C and growth rates were observed for 15 d.

RESULTS AND DISCUSSION

Chilling injury during cold storage

Chilling injury symptoms of fruits stored at 0°C were found from about day 15 of storage. Some of stored peppers showed a change in calyx color from green to dark green and water-soaking (Fig. 1A). The brown seed rate was about 2-fold higher in cold-treated peppers compared to MeJA-, 1-MCP-treated peppers in 15 days after storage. And approximately 30% higher seed browning were represented in cold-treated peppers compared to MeJA-, 1-MCP-treated peppers in 20 days after storage (Fig. 1B). Sheet pitting was also showed about 2-fold highly in cold-treated peppers compared to MeJA-, 1-MCP-treated peppers during the period for investigation (Fig. 1C). From these results, it was confirmed that MeJA and 1-MCP have a function to alleviate chilling injury of pepper fruits as reducing seed browning and sheet pitting corresponding to previous reports which demonstrate stress defense reaction of MeJA and 1-MCP (Anderson et al. 2004; Fung et al. 2004; Cao et al. 2009).

IDENTIFICATION OF DIFFERENTIALLY *EXPRESSED* GENES IN COLD STORAGE

A total of 778 genes were differentially expressed (more than 2 fold) in three groups of cold treated peppers compared to non-treated peppers . Therein, 424, 761 and 509 genes corresponded to cold-, MeJA- and 1-MCP-treated peppers, respectively, and 380 genes were common to each treatment. Of these, 240 genes were upregulated and concurrently 84 genes were downregulated in cold-treated pepper compared to non-treated peppers. And 470 genes were upregulated and 291 genes were downregulated in MeJA-treated peppers, 290 genes were upregulated and 219 genes were downregulated in 1-MCP-treated peppers, respectively (Table 1).

To examine the genes represented expression change under the influence of cold and the candidate genes conferring tolerance to cold, differentially expressed genes in either MeJA- or 1-MCP-treated pepper, which showed simultaneously significant expression change in cold-treated peppers, became an object of functional enrichment analysis . All of genes in response to both cold and substance were characterized functionally using GO term based on biological function .106 genes analyzed by GO term among total 423 genes were evaluated enrichment. GO analysis represented significant enrichment such as genes relative to stress response, hormone stimulus containing ethylene signaling and secondary metabolic process. Of these, they were characterized that 29 genes were relative to stress response, 14 genes were relative to hormone stimulus and 10 genes were concerned with secondary metabolic process.

1. The genes relative to stress response. 29 genes characterized this category were as follow. the genes concerned with defense reaction such as WRKY, allene oxide synthase, lipoxygenas and phenylalanine ammonia-lyase; especially cold response such as sulfite reductase, catalase, salt tolerance zinc finger domain and 3-ketoacyl-coA sythase. MeJA treatment highly upregulated genes for peroxidase and allene oxide synthase, compounds related to stress responses (Table 2).

2. The genes relative to hormone stimulus. 14 genes characterized this category were as follow. 6 genes such as ethylene responsive element binding factor, MAP kinase kinase and ethylene insensitive were relative to ethylene mediated signaling pathway; numerous genes containing MYB domain, MYC domain and auxin-responsive protein were affiliated to this category.

3. The genes relative to secondary metabolic process. 14 genes were characterized this category such as gibberellin 2-oxidase, cinnamate-4-hydroxylase and 9-cis-epoxycarotenoid dehydrogenase.

4. Transcription factors significantly differential expressed. 17 of transcription factors such as bHLH-type transcription factor, AP2 domain and TBP-associated factor showed notably expression change.

Regulation and integration of ERFs in MeJA and ethylene signaling pathway conferring to stress tolerance

To clarify the role of crosstalk between ethylene and MeJA in regulating chilling stress, *ERF* expression in harvested peppers was examined during the long-term, low-temperature storage. Transcriptome and qRT-PCR analysis showed that *ERF* expression was upregulated by MeJA under cold temperature (Fig. 2). *ERF1*, *ERF2*, and *ERF5* were markedly upregulated by MeJA compared with cold treatment alone. MeJA-treated peppers showed higher expression levels of *ERF1*, *ERF2*, and *ERF5* than did the cold-treated peppers for 14 days of cold storage at 0°C (Fig. 2). 1-MCP treatment inhibited *ERF* expression, whereas an increase in the duration of cold storage at 0°C increased *ERF* expression. Accumulation of cold stress also induced *ERF* expression. Upon returning to 20°C, the levels of ERF transcripts decreased dramatically on day 17 in fruits treated with MeJA, 1-MCP, and cold. Among the ERFs, ERF1 showed a more sensitive response to MeJA than did ERF2 and ERF5, mainly because ERF1 expression levels were higher than those were of ERF2 and ERF5. Additionally, increased *ERF1* transcript levels correlated with reduced
Transformed *E. coli* cells overexpressing NAC, WRKY and ERFs were more tolerant to cold condition representing that advanced growth rate was shown in *E. coli* overexpressing NAC, WRKY and ERFs compared to transformed *E.coli* overexpressing empty vector when grown in cold condition (Fig.3A). Furthermore, redox status of transformed *E.coli* exposed to stress conditions was examined (Fig. 3B). ROS contents were remarkably alleviated (20-30% lower) in *E. coli* overexpressing NAC, WRKY and ERFs compared to *E.coli* overexpressing empty vector with exposure to cold condition (0°C) for 7 days. In addition, H_2O_2, the direct source of ROS, was treated as stressor to reconfirm the function of genes scavenging ROS effectively. Transformed *E. coli* overexpressing NAC, WRKY and ERFs also showed lower levels of ROS (20-40%) compared to Transformed *E.coli* overexpressing empty vector.

In conclusion, we investigated global expression changes in peppers against cold storage and substance treatment, MeJA and 1-MCP, which improved tolerance to cold condition. The expression of 778 transcripts showed significant alteration more than 2-fold by microarray analysis. These results suggested that MeJA conferred tolerance to peppers against chilling injury by activating genes relative to stress response, ethylene signaling and various transcription factors. Numerous genes were related simultaneously to enriched categories, and it supported that these genes involved in crosstalk among various reactions to improve cold tolerance. We thought that alleviation of chilling injury by MeJA was facilitated as activating antioxidant and stress response genes concurrently induced by MeJA response, as confirming that chilling injury occurred by oxidative stress. It was predicted that MeJA regulated ethylene induced genes under cold storage; effectively regulation of transcription factors like ERFs, key elements in integration of jasmonate and ethylene signals for the regulation of defense response genes.

LITERATURE CITED

Anderson JP, Badruzsaufari E, Schenk PM, Manners JM, Desmond OJ, Ehlert C, Maclean DJ, Ebert PR, Kazan K 2004. Antagonistic interaction between abscisic acid and jasmonate-ethylene signaling pathways modulates defense gene expression and disease resistance in Arabidopsis. Plant Cell 16:3460-3479

Cao S, Zheng Y, Wang K, Rui H, Tang S 2009. Effects of 1-methylcyclopropene on oxidative damage, phospholipases and chilling injury in loquat fruit. J. Sci. Food Agric. 89:2214-2220

Collén J, Hervé C, Guisle-Marsollier I, Léger JJ, Boyen C 2006. Expression profiling of Chondrus crispus (Rhodophyta) after exposure to methyl jasmonate. J. Exp. Bot. 57:3869-3881

Fung RW, Wang CY, Smith DL, Gross KC, Tian M 2004. MeSA and MeJA increase steady-state transcript levels of alternative oxidase and resistance against chilling injury in sweet peppers (*Capsicum annuum* L.). Plant Sci. 166:711-719

Larrigaudiere C, Vilaplana R, Soria Y, Recasens I 2004. Oxidative behaviour of Blanquilla pears treated with 1-methylcyclopropene during cold storage. J. Sci. Food Agric. 84:1871-1877

McDowell JM, Dangl JL (2000) Signal transduction in the plant immune response. Trends Biochem Sci 25: 79-82

Özden Ç, Bayindirli L 2002. Effects of combinational use of controlled atmosphere, cold storage and edible coating applications on shelf life and quality attributes of green peppers. Eur. Food Res. Technol. 214:320-326

Parra-Lobato MC, Fernandez-Garcia N, Olmos E, Alvarez-Tinaut MC, Gómez-Jiménez MC 2009. Methyl jasmonate-induced antioxidant defence in root apoplast from sunflower seedlings. Environ. Exp. Bot. 66:9-17

Ryall AL, Pentzer WT 1982. Handling, transportation and storage of fruits and vegetables. Volume 2. Fruits and tree nuts. AVI Publishing Co., Inc.,

Sasaki-Sekimoto Y, Taki N, Obayashi T, Aono M, Matsumoto F, Sakurai N, Suzuki H, Hirai MY, Noji M, Saito K 2005. Coordinated activation of metabolic pathways for antioxidants and defence compounds by jasmonates and their roles in stress tolerance in Arabidopsis. Plant J. 44:653-668

Sasaki Y, Asamizu E, Shibata D, Nakamura Y, Kaneko T, Awai K, Amagai M, Kuwata C, Tsugane T, Masuda T 2001. Monitoring of methyl jasmonate-responsive genes in Arabidopsis by cDNA macroarray: self-activation of jasmonic acid biosynthesis and crosstalk with other phytohormone signaling pathways. DNA Res. 8:153-161

Sisler E, Serek M 1997. Inhibitors of ethylene responses in plants at the receptor level: recent developments. Physiol Plant 100:577-582

Wang K, Jin P, Cao S, Shang H, Yang Z, Zheng Y 2009. Methyl jasmonate reduces decay and enhances antioxidant capacity in Chinese bayberries. J. Agric. Food Chem. 57:5809-5815

Yuan G, Sun B, Yuan J, Wang Q 2010. Effect of 1-methylcyclopropene on shelf life, visual quality, antioxidant enzymes and health-promoting compounds in broccoli florets. Food Chem. 118:774-781

Ziliotto F, Begheldo M, Rasori A, Bonghi C, Tonutti P 2008. Transcriptome profiling of ripening nectarine (*Prunus persica* L. Batsch) fruit treated with 1-MCP. J. Exp. Bot. 59:2781-2791

Table 1. The number of genes with significant differential expression in peppers during cold storage, determined by microarray analysis.

Sample	Up regulation	Down regulation	Total significant genes
Significantly changed genes			
Cold	240	184	424
MeJA	470	291	761
1-MCP	290	219	509

Table 2. The list of genes highly up regulated by MeJA treatment compared to cold treatment

	Fold change			
SEQ_ID	Cold	MeJA	1-MCP	Description
PEPPERS0013794	2.43	7.48	1.68	MYB77
PEPPERS0002364	2.94	4.60	3.38	WRKY40
PEPPERS0003048	7.38	13.41	12.19	ERF2
PEPPERS0012323	2.76	11.79	2.92	catalase 2*
PEPPERS0010090	5.52	8.10	5.50	MKK9 (MAP kinase kinase 9)
PEPPERS0000100	3.40	9.33	4.44	RHD2 (root hair defective 2)
PEPPERS0009982	16.03	20.57	13.68	STZ (salt tolerance zinc finger)*
PEPPERS0000403	3.23	16.52	3.22	NAC domain
PEPPERS0000630	2.21	4.38	2.09	trans-cinnamate 4-monooxygenase
PEPPERS0018525	3.00	7.90	1.80	ERF1 (ethylene responsive factor 1)
PEPPERS0013782	3.78	6.92	4.47	RD21 (responsive to dehydration 21)
PEPPERS0002578	2.83	30.24	4.67	JAZ1 (jasmonate-ZIM-domain protein 1)
PEPPERS0003831	7.00	64.01	3.96	peroxidase
PEPPERS0009632	9.03	10.81	7.13	ERF5
PEPPERS0018370	2.48	19.78	4.82	AOS (allene oxide synthase)
PEPPERS0002556	2.95	13.42	4.87	lipoxygenase

Figures

A.

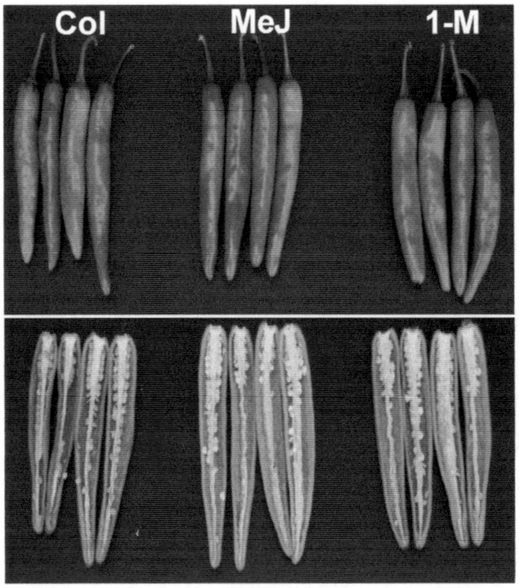

B. C.

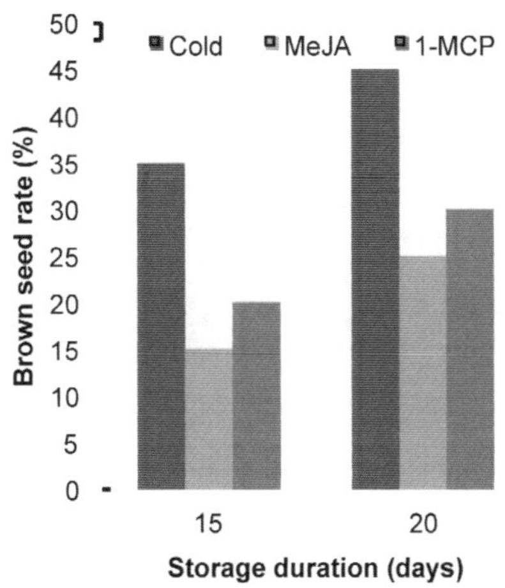

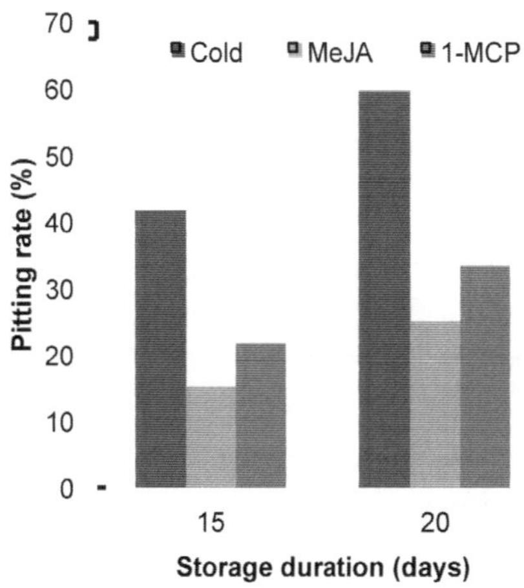

Figure 1. Seed browning and pitting rate of peppers stored at 0°C. A, phenotypes of pepper stored at 0°C for 20 days. B and C, Seed browning rate (B) and pitting rate (C) of peppers stored at 0°C for 15 days and 20 days. Cold, peppers stored in cold condition (0°C); MeJA, peppers treated with 50 μM MeJA and then stored at 0°C; 1-MCP, peppers treated with 1 ppm 1-MCP and then stored at 0°C.

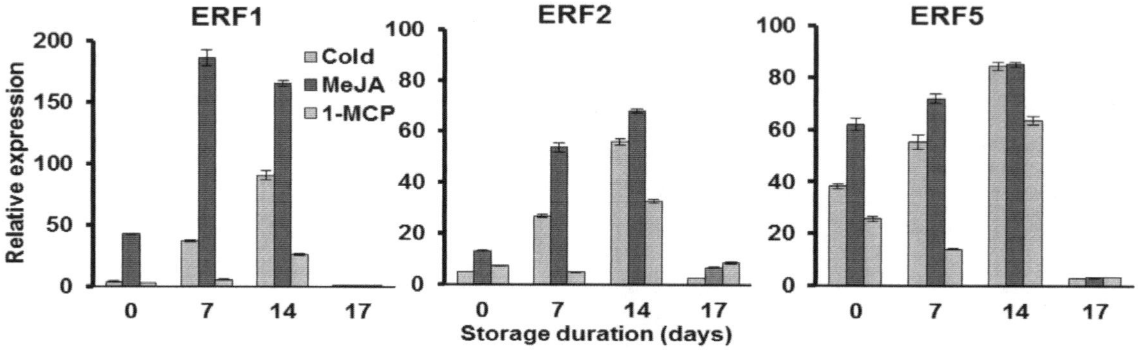

Figure 2. Abundance of *ERF* gene transcripts for cold tolerance in peppers analyzed by qRT-PCR. The transcript levels are relative to the expression levels of genes in non-treated peppers. Freshly harvested peppers were treated at 25°C for 24 h with 50 μm methyl jasmonate (MeJA), 1 ppm 1-MCP (1-MCP), and air (Cold). After the treatment, the peppers were stored at 0°C for 14 days (as analyzed on days 0, 7, and 14) and were rewarmed at 20°C for another 3 days (17 days in total). The results represent mean ± SD of 3 independent experiments.

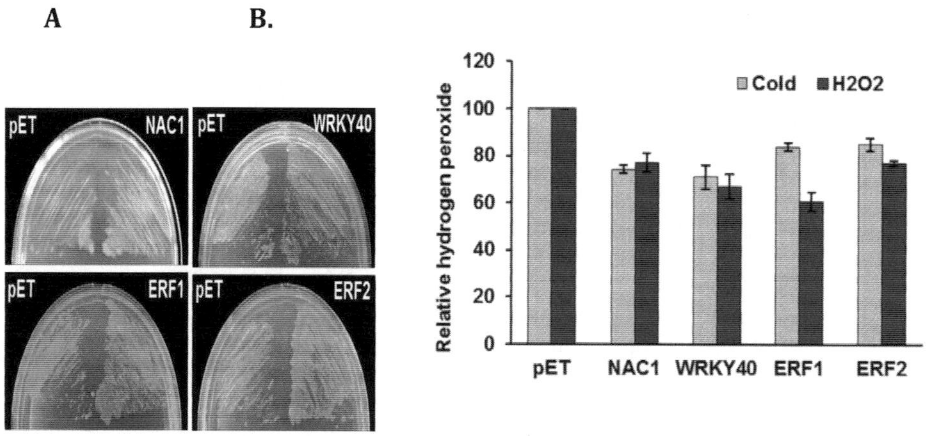

Figure 3. Stress tolerance and redox status of transformed *E. coli* overexpressing *NAC, WRKY,* and *ERF* genes. A Growth rate of transformed *E. coli* overexpressing pepper genes grown at 7°C, monitored for 13 days.. B, Redox status of transformed *E. coli* in the presence of cold and oxidative stressors. The level of hydrogen peroxide in *E. coli* in response to cold (0°C) and 1.5 mM H_2O_2 was determined using FOX reagent. Oxidation of ferrous ion indicated by xylenol orange was measured pectrophotometrically. The means ± SD of three independent determinations are shown.

'CHOCOLATE BEAUTY' BELL PEPPER RESPONSE TO FOLIAR BIOSTIMULANT APPLICATIONS AND FERTILIZATION METHODS

Pedro E. Cruz-Carballo, J. Pablo Morales-Payan, Feiko Ferwerda, Dania Rivera-Ocasio

University of Puerto Rico, Mayaguez Campus, Mayaguez, PR

An experiment was conducted in July-November 2014 in Mayaguez, Puerto Rico, to determine the effects of biostimulants and fertilization on growth, yield, and quality of 'Chocolate Beauty' bell pepper grown in containers in a high tunnel protective structure. The plants were grown in a 3:1 mixture of alluvial soil and sphagnum + perlite mix (Promix® BX), fertilized with either (1) 22.67 kg N-P-K/0.40 ha pre-plant and 45.35 kg N/0.40 ha 10 weeks later using granular urea (=field fertilization recommendation by Ag Experiment Station, UPRM), (2) 22.67 kg N-P-K/0.40 ha pre-plant and 45.35 kg N/0.40 ha divided in 10 applications every 7 days using an urea solution (=greenhouse fertilization), and (3) 22.67 kg N-P-K/0.40 ha from organic 6-6-5-8Ca + 0-0-15-1Ca pre-plant and 45.35 kg N/0.40 ha from organic 6-0-0-8Ca (Bioflora®). Every 14 days the crop leaves were sprayed to run-off with different biostimulant solutions. The biostimulants tried were (1) a blend of amino acids formulated for crops (AA) (Aminoquelant Ca®), at 4 ml/L of water, (2) an extract of the marine alga Ascophyllum nodosum formulated for crops (Stimplex®), at 5 ml/L, (3) a mixture of natural ingredients containing brassinosteroids, vitamins, and enzymes (Vitazyme®), at 10 ml/L of water, and (4) water (control). Fruits were harvested at maturity (color turning from green to chocolate), weighted, quantified, and graded into categories according to their diameter and quality. Total fruit number and weight per plant were affected only by fertilizers, with the soluble treatment resulting in more fruits per plant than the other treatments. Marketable fruit number was significantly lower in plants treated with granular fertilizer. There were no significant effects of treatments on shoot fresh weight, root fresh weight, root dry weight. Plants treated with dissolved fertilizer accumulated more shoot dry weight and fruit dry weight than plants treated with granular fertilizer. SPAD values (directly related to chlorophyll concentration) on the most recently expanded leaves were significantly higher in plants receiving organic or dissolved fertilizer than in plants treated with granular fertilizer. There were significant interactions of biostimulants and fertilizers on the concentration of soluble solids in the marketable fruits, with the highest values found in fruits from plants treated with organic fertilizer and the vitamin-based biostimulant, while the lowest values (30% lower) were found when plants were treated with no biostimulant and dissolved fertilizer. These results indicate that the choice of fertilizer may significantly impact the growth of the specialty bell pepper used in this research, while biostimulants may be used to enhance fruit quality by increasing the concentration of soluble solids in the fruit. Disclaimer: Mentioning brand names does not imply an endorsement from the authors or from the University, and is done solely to specify the materials used in this research.

EFFICACY OF METAMITRON AS A POSTBLOOM THINNER – THE BRAZILIAN EXPERIENCE

P. Francescatto, T. L. Robinson, Cornell University, Geneva, NY, USA

A. Rufato, Embrapa Vacaria, RS, Brazil,

L. Rufato, J.F. Carminatti, University of the Santa Catarina State, Lages, SC, Brazil

Climate conditions of Southern Brazil have been largely considered one of the main factors responsible for the year to year variation in fruit set of apples. Blossom thinning becomes a very risk decision for growers; and intensity and need of chemical thinning is often decided after flower fecundation. Since the use of Carbaryl was banned in Brazil, the range of thinners became narrow and the search of new alternatives has been priority. Therefore, the main goal of this experiment was to study the effect of timing and rate of metamitron on apple fruit thinning related to the carbohydrate balance predicted by the Malusim Model. The trial was carried out in the Southern area of Brazil - Vacaria, RS, during the crop season 2014/15. The cultivar tested was 'Gala' trained as slender spindle and planted at 2353 trees/ha. Three rates of metamitron were tested: 100, 200 and 300 ppm sprayed four times from petal fall to 18-20mm fruit diameter at seven days intervals. A parallel trial was conducted to test the combination of 60 ppm of BA plus 100 or 200 ppm of metamitron in order to potentially improve thinning efficacy and fruit size. Metamitron 300 or 200 ppm applied at petal fall or 300 ppm 21 days after petal fall (18-20mm fruit size) and Metamitron 200 ppm + BA at petal fall over-thinned the trees. According to the model, there was a considerable carbohydrate deficit during those two periods. Regardless metamitron rates or BA addition, no effect on fruit thinning was observed between 8 to 12 mm fruit diameter. The higher the effect of metamitron on fruit thinning the larger was the fruit. Addition of BA to metamitron seemed to improve the efficacy for thinning. However, fruit size was not increased by BA in a tank mix with metamitron; higher rates of BA and low rates of metamitron should be tested. Metamitron 100 ppm + BA 60ppm seemed to adjust the crop at appropriate levels under periods of low carbohydrate balance. In general, metamitron showed efficacy for Gala apple thinning. Metamitron responses in 'Gala' could be greatly correlated with the carbohydrate balance observed for this location.

RESEARCH PROGRESSES OF PHYTOHORMONAL QUANTIFICATION BASED ON MOLECULAR METHODS

Ruozhong Wang, Lixiong He, Zhigang Huang, Langtao Xiao

Hunan Provincial Key Laboratory of Phytohormones and Growth Development, Hunan Agricultural University, Changsha, China; Hunan Co-Innovation Center for Utilization of Botanical Functional Ingredients, Changsha, China

Molecular methods have been employed in phytohormonal assay since the early 1990s, successful examples include the assay based on the IAA and CTK inducible Pg5::GUS expression, the use of DR5-based auxin inducible reporters to monitor auxin distribution and the recent application of the phytohormonal degrons, i.e. the fast degradation of repressor proteins triggered by phytohormone perception. Generally, molecular methods employed in phytohormonal quantification include assays based on phytohormonal action related events and phytohormonal signal transduction pathways, as well as other assays using molecular techniques. In recent years, several phytohormonal quantification techniques based on molecular methods have been established in the authors' laboratory. These techniques include: a) The terminal protection of phytohormone linked single strand DNA (ssDNA) and double strand (dsDNA) based on the stereo-hindrance effect caused by molecule recognition, which showed limits of detection (LODs) at pmol level for several phytohormones; b) The immuno-PCR based on the polymer chain reaction of phytohormone conjugated DNA strands, which obtained LODs at sub-nanogram level; c) Förster resonance energy transfer (FRET) biosensors based on the fluorescence wavelength and intensity shift caused a phytohormone conjugated ssDNA, which had LODs at sub-nonograms; d) Promoter-reporter and degron-reporter systems for several phytohormones, which showed the potential for in vivo and cell specific quantification.

STUDY OF THE EFFECT OF NATURAL ROOT ACCELERATORS ON PLANT ROOTING

Konstantin Chakalov[1], Valentin Savov[2], Gergana Angelova[1], and Todorka Popova[1]

E-mail: konstantinchakalov@yahoo.com

[1] Balkan Plant Sciences, 34, Zlatovruh Street, floor 1, Sofia 1164, Bulgaria
[2] Sofia University "St.Kliment Ohridski", 8 "Dragan Tzankov" Blvd., 1164 Sofia, Bulgaria

The presence of phytohormones and their analogues in brown, red, green and blue-green algae has been found long ago. The aim of this study is to determine the effect of extracts from micro and macro algae on the process of plant rooting.

The influence of extracts from macroalgae: Ecklonia maxima, Ascophyllum nodosum, Durvillaea potatorum, Laminaria sp. and microalgae: Chlorella sp., Scenedesmus sp, Spirulina sp. Individually or in combination with other natural extracts is tested on cascade petunia and chrysanthemum grown in peat-perlite substrate.

Phytohormones; Standard error *± 3,1 % RSD*	Rhizotonic®	HTSW	*Cetraria islandica*	*A. nodosum*	*D. potatorum*	*Chlorella sp,*	*Spirulina sp,*
	ng/ml						
GA-1	n,d,	n,d	1,68	n,d,	n,d,	n,d,	n,d,
GA-3	0,93	n,d	2,15	2,765	n,d,	n,d,	n,d,
GA-4	2,95	4,75	0,53	5,504	n,d,	0,696	0,416
GA-5	17,20	n,d,	2,32	n,d,	n,d,	n,d,	n,d,
GA-6	2,91	n,d	0,67	n,d,	n,d,	n,d,	n,d,
GA-7	10,55	5,15	n,d	4,125	n,d,	n,d,	n,d,
GA-8	n,d,	n,d	9,34	n,d,	n,d,	n,d,	n,d,
GA-9	2,95	n,d	0,45	n,d,	n,d,	n,d,	n,d,
trans-Zeatin	5,50	13,65	22,68	7,892	3,02	162,835	17,982
CPPU,4-CPPU	2,29	3,46	1,05	4,065	0,13	0,439	0,183
N-Benzyladenin (BA)	6,61	n,d	124,10	n,d	n,d	n,d	n,d
IAA	2,77	n,d,	4,86	n,d,	n,d,	n,d,	n,d,
IPA	3,02	4,03	0,49	10,285	14,03	5,664	8,552
dCPAA	10,74	6,10	1,54	3,618	n,d,	1,405	0,954
IBA	2,06	2,20	0,53	3,806	0,35	n,d,	n,d,
Thidiazuron	1,37	3,66	0,72	5,258	n,d,	n,d,	n,d,
Abscisic acid	5,34	2,56	11,09	4,821	34,46	1,458	1,810
Ancymidol	1,23	n,d	0,15	6,738	n,d,	3,520	1,543
Paclobutrazol	2,12	3,08	0,51	5,335	0,14	0,289	0,050
Fluprimidol	3,19	4,04	n,d	4,290	0,04	0,103	0,123
Uniconazol	2,18	4,47	0,53	3,711	0,12	0,146	0,296

Table 1. Content of some phytohormones in the studied extracts

The plants from each variant are treated once with a 2 mL/L solution of the algal extracts on the background of hydroponic nutrient solution Grow, Micro, Bloom (Advanced Nutrients Ltd., Canada). The root biomass is measured on the 11th day after transplanting.

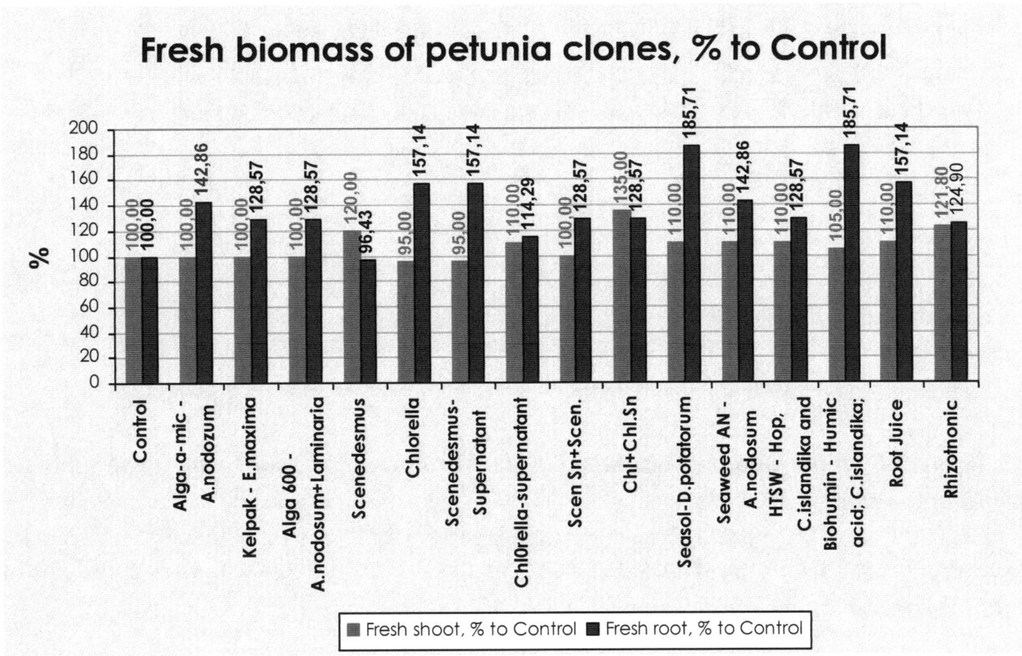

Fig. 1. Influence of the studied extracts on the formation of shoot and root biomass of petunia clones

In the test with petunia the most profound impact on rooting has the extract from D. potatorum - the formed root biomass is 85.7% heavier than that of the Control. The E. maxima extract increases the root biomass by 28.57 % and the A. nodosum extracts – by 28.57 and 42.86% (Fig. 1).

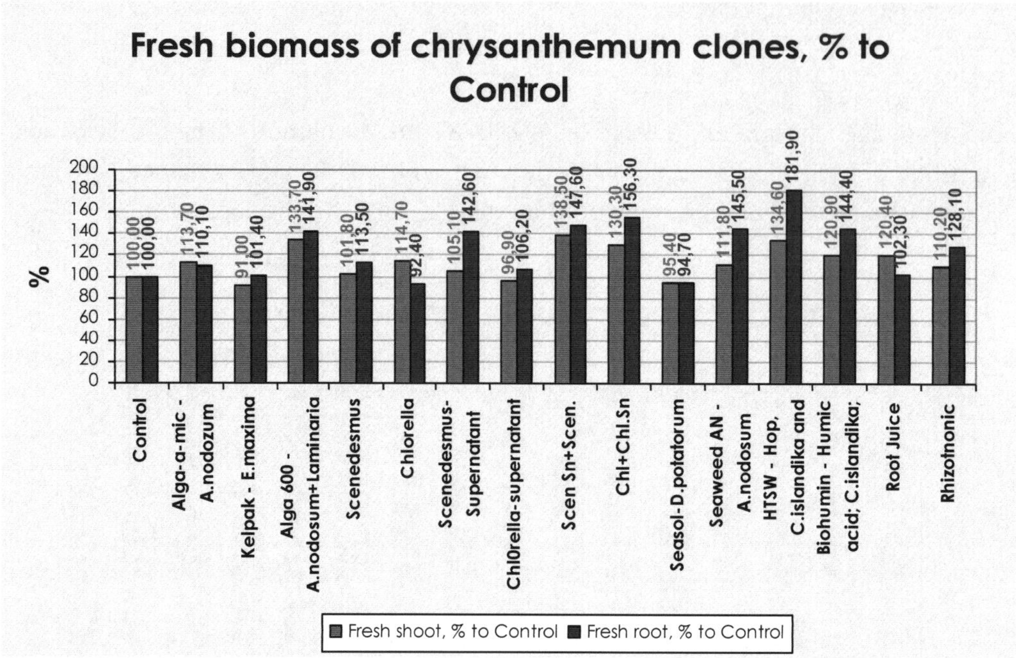

Fig. 2. Influence of the studied extracts on the formation of shoot and root biomass of chrysanthemum clones

In the case of chrysanthemum treatment, best results are obtained using the A. nodosum extracts - the root biomass increase is 41.9 and 45.5% compared to the Control (Fig. 2).

Our modifications HTSW and Biohumin used for comparison have a very strong impact on the formation of the root system of both types of test plants, increasing the root biomass by up to 81.9 and 85.71% respectively. The treatment with micro algae shows that the application of Chlorella (cell suspension or supernatant) increases the root biomass up to 57.14% (petunia) and 56.3% (chrysanthemum). The commercial products Rhizotonic and Root Juice have lower effect on the rooting of the test plants than the tested algal extracts and their combination with other natural substances.

Conclusions:

- The ratio auxines to cytokinins in the algal extracts is more important for their application as Root Accelerators than the total content of these hormones.
- In this test kelp *A. nodosum* extracts have more consistent effect than the other kelps.
- Microalgal extracts can be used as an additive for improving the composition of other extracts for root acceleration

STS APPLIED INSIDE FOLIAR ROSETE DOES NOT INHIBIT THE PINEAPPLE NATURAL FLOWERING

Victor M. Maia ,Virgínia D. Araújo, Lucas B. Ferreira, Rodinei F. Pegoraro, Ignácio Aspiazú, Josimara M. Rabelo

State University of Montes Claros

Brazil is the second largest producer of pineapple (Ananas comosus var. Comosus), with a planted area of 62,481 hectares and production of 2.3 million tons. Natural flowering of pineapple is a complex process with multifactorial control and presents several drawbacks for directly affecting the management and conduction of the crop and the harvesting and marketing of fruits. Therefore, the aim of the study was to determine the effect of silver thiosulfate (STS) in inhibiting natural pineapple flowering. The experimental design was a randomized block in a factorial 4 x 5, for 4 cultivars and 5 STS doses, with 4 replicates being each pot an experimental plot. The cultivars studied were 'IAC Fantastic','Pérola', 'MD2' and 'Vitória'. STS doses were 0; 0.5; 1; 2 and 3 mM STS that were applied inside the leaf rosette pineapple, twice a month with an interval of 15 days, during June and July. The characteristics evaluated were the percentage of plants with natural flowering, fruit weight with and without crown, fruit length, fruit diameter and number of seedlings per plant. The soluble solids (ST), texture, hydrogen potential (pH) and titulable acidez (AT) of the pulp also were determined. Beyond that, the fresh and dry matter weight of each compartment of the plant were determined. The results showed that application of STS inside the rosette doesn't inhibit the natural pineapple flowering, increases the diameter of the fruit and leaf dry matter 'D', and reduce the pH of the pulp. Support: Fapemig, CNPq and CAPES.

THE ROLE OF ROOT ARCHITECTURE IN PLANT ACQUISITION OF WATER AND MINERAL NUTRIENTS UNDER LIMITING CONDITIONS

Leon V. Kochian

USDA-ARS Robert W. Holley Center for Agriculture and Health at Cornell University

Plant root systems function as critical links between the growing shoot and the rhizosphere, providing both nutrients and water. In recent years, agricultural researchers have focused an increased amount of attention on root system architecture (RSA), as it has been shown that plants exert a significant degree of genetic control over this complex trait. Furthermore, there is an increasing awareness that RSA plays an important role in crop nutrient acquisition efficiency. Differences in RSA can lead to root archetypes that are better at acquiring nutrients such as phosphorous that have relatively low mobility in the soil, as well as root architectures better designed to acquire mobile soil nutrients such as nitrate and water. In recent years in an effort to address the major bottleneck in root architecture research – phenotyping of RSA with high enough throughput and resolution to genetically analyze RSA traits, a number of new root imaging/analysis platforms have been developed. In this presentation, a root system architecture phenotyping and analysis platform is described that was designed to improve the flexibility and throughput for root system phenotyping. The platform employs digital imaging and software analysis tools to quantify root system architecture (RSA) in 3-dimensions (3D). Our initial work on rice employed plants whose roots were grown in transparent gellan gum with nutrients. Subsequently, we moved on to specially constructed plastic mesh systems that maintain 3D RSA but do not impede root growth, and tis system is being used to phenotype root systems in hydroponics and soil-like media. The use of the 3D phenotyping platform was used to quantify RSA traits for roots of rice and sorghum plants from linkage and association populations, and quantitative trait loci (QTL) and genome wide association (GWA) mapping studies were performed in order to investigate the genetic components of RSA development. The RSA analysis of lines from the sorghum association panel was used along with field phenotyping of the same panel for P efficiency traits to investigate the role of sorghum homologs of the rice P efficiency gene, OsPSTOL1, in sorghum P efficiency.

POST-HARVEST BIOLOGY OF PLANT-PATHOGEN INTERACTIONS: TRANSCRIPTIONAL HIJACKING OF PLANT STRESS HORMONES BY BOTRYTIS CINEREA INFECTION

Estefanía Vincenti[1], Barbara Blanco-Ulate[2], Ann L.T. Powell[1] and Dario Cantu[2]

[1]*Department of Plant Sciences, University of California, Davis, CA, USA.*

[2]*Department of Viticulture and Enology, University of California, Davis, CA, USA*

Plant hormones act as signals to provide external and internal information to cells throughout the plant. Generally, they are present in low concentrations and mediate multiple adaptive responses. Changes in hormone concentration or sensitivity are essential for the regulation of plant growth, development, reproduction and responses to biotic and abiotic stresses (Hooykaas et al., 1999; Crozier et al., 2000; Davies, 2004). The nature of hormone regulation has been described as a complex network of cross-communicating signaling pathways (Pieterse et al., 2009) that involve several hormones with major or minor responsibilities depending on the process and the environment. Key roles in the induction and modulation of plant's immune response by acting as the primary signally molecules in plant-pathogen interactions have been described for ethylene (ET), salicylic acid (SA) and jasmonic acid (JA) and, recently, abscisic acid (ABA). Most of the roles characterized for plant-stress hormones have been studied in vegetative tissues (Wasternack, 2007; Bari and Jones, 2009; Pieterse et al., 2009; Cutler, 2010; Rivas-San Vicente and Plasencia, 2011; Vandenbussche and Van Der Straeten, 2012; Pieterse et al., 2012), thus we have a limited understanding of these hormones' role in the regulation of fruit-fungal infections.

Tomato (*Solanum lycopersicum*) and *Botrytis cinerea* present an ideal pathosystem to elucidate the molecular conversation involved in fruit defense responses and fungal colonization strategies. The necrotrophic fungus *B. cinerea* is unable to colonize unripe (Mature Green; MG) tomato fruit, but it grows extensively on ripe (Red Ripe; RR) fruit. Increased susceptibility has long been observed in the later stages of tomato fruit ripening, assumed to be an inherent consequence of the ripening process. Our research group integrated gene expression analysis with susceptibility assays of tomato fruit compromised in hormone synthesis and/or perception; to hypothesize how ET, SA, JA and ABA influence the resistance or susceptibility of unripe and ripe tomato fruit to *B. cinerea*.

ET production during infection represents an early response of plants to the perception of a pathogen attack and can be associated with activation of plant defense-related processes such as the production of phytoalexins, PR proteins, the induction of the phenylpropanoid pathway and cell wall alterations (Rodrigo et al., 1993; Fan et al., 2000; van Kan, 2006). Conversely, ET is also considered very important in the development of disease symptoms (Boller, 1991; Abeles et al., 1992, Cristescu et al., 2002). This hormone was considered a sensitive marker for early infection of harvested fresh products, as a significantly higher level of ET production was demonstrated in *B. cinerea*-infected tomatoes than the level in wounded but not inoculated ones. It was described that ET concentration began to increase before there was visible decay (Polevaya et al., 2002). Thus, the presence of ET may provide an advantage to the fungus indirectly, because it stimulates the softening of the plant tissue and therefore facilitates tissue penetration and fungal colonization (Diaz et al., 2002). In addition, as *B. cinerea* can synthesize ET and the analysis of its emission from a host-pathogen system is a complex problem, many biological questions remain regarding the role of ET in pathogenesis and resistance.

METHODOLOGY

From the AHD 2.0 (Arabidopsis Hormone Database), Blanco-Ulate et al. (2013) identified genes previously implicated in the biosynthesis, signaling and response pathways of ET, SA, JA and ABA. The retrieved sequences were blasted (BLASTP) against the tomato predicted proteome, to identify putative tomato homologs to the *Arabidopsis* genes using an identity greater than 60% and alignment coverage over 70% of the query length as selection parameters. The resulting candidate gene list was complemented with known candidates related to ET pathways in tomato fruit. Candidate genes were searched in the microarray dataset provided by Cantu et al., (2009), which comprises transcriptional changes of ripening-related genes in MG and RR tomato fruit due to infection of *B. cinerea* at 1 day post infection. Significant ($P \leq 0.05$) transcriptional fold changes were identified in genes that were either uniquely present in infection or due to healthy ripening. A total of 65 candidate genes were identified and are detailed in Blanco-Ulate et al. (2013)

RESULTS & DISCUSSION

Ethylene Biosynthesis Gene Expression Analyses

The expression patterns of *LeSAMS3* and *LeACO5,* which encode for enzymes in the biosynthetic pathway of ET, are shown in figure 1. In infected MG tissues, *LeSAMS3* shows a significant up-regulation at both 1 and 3dpi, whereas for *LeACO5* expression, it is suppressed at 1dpi, but induced at 3dpi; however, the fold change is only significant for the first time point. There are no significant changes in *LeSAMS3* expression during ripening or infections of RR fruit at 1dpi, although a significant down-regulation is observed at 3dpi. No significant change in expression is observed for *LeSAMS3* during ripening. Conversely, as fruit ripen normally (purple bar) at both 1 and 3 d, the transcript abundance of *LeACO5* is significantly increased.

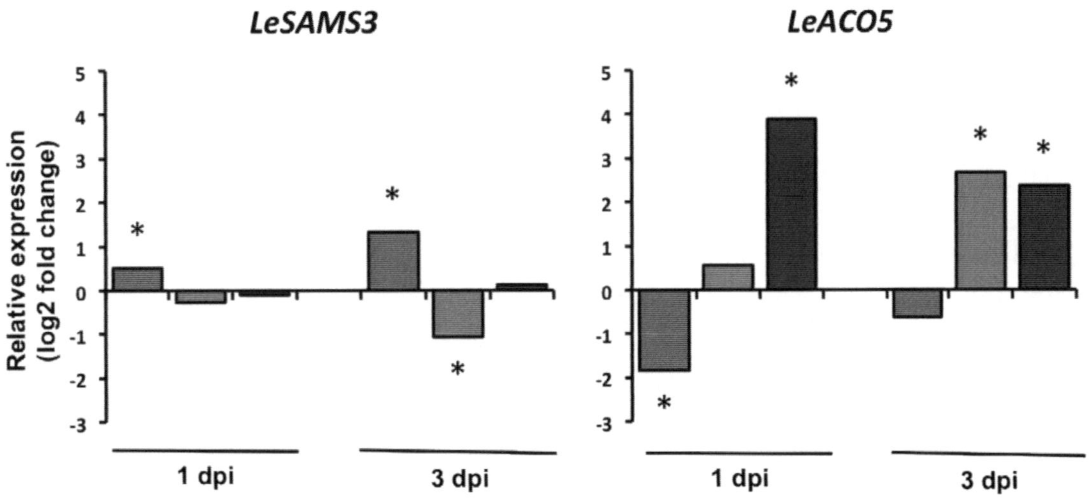

Figure 1. Changes in expression profiles of ET-biosynthetic genes *LeSAMS3* and *LeACO5* at 1 and 3 dpi due to infection by B. cinerea (in MG or RR) and ripening; significant changes (*P value* ≤ 0.05) are marked with an asterisk on top of below the bars.

The expression of *LeSAMS3*, which converts methionine into SAM, is crucial in activating the pathway of ET production. Results obtained by van de Poel et al. (2012) showed that of the four SAMS genes (*SAMS1–SAMS4*), SAMS3 was implicated in fruit development and in general none were expressed at high levels during ripening, although SAM levels peaked during these stages. This is consistent with the data shown in the purple bars that represent the comparison between RR and MG healthy tissue, in which no significant change is observed at either time points. ET and polyamines (PAs) biosynthetic pathways share the common intermediate SAM. Polyamines (e.g., spermidine (Spd) and spermine (Spm)) are compounds implicated in multiple growth and developmental processes, such as the inhibition of senescence (Mehta et al. 2002; Nambeesan et al. 2010, Bitrián et al. 2012). Spd and Spm are synthetized from dcSAM, to Ptc or Spd, respectively by S-adenosylmethionine decarboxylase (SAMDC, i.e., homolog to the plants' SAMS) (Bagni, 1989). SAM is both the substrate for ACS in ET biosynthesis and for SAMDC in PA biosynthesis.

As MG tissues are usually resistant to *B. cinerea* infection, the up-regulation of *LeSAMS3* at both 1 and 3 dpi is likely to increase SAM synthesis and alter both ET and PA production. It was suggested by Tassoni et al., (2006) that PAs could modulate the rate of ripening and over-ripening by counteracting the ET-triggered ripening responses. The results in figure 1 suggest that these changes in the relative synthesis of ET/PA may be a control point for unripe fruit to regulate ET synthesis, delaying ripening and ultimately promoting resistance (Mehta et al. 2002; Tassoni et al., 2006; Nambeesan et al. 2010; van de Poel et al. 2012).

This defense mechanism may be implicated in response to infection in early stages of ripening, but once ET levels have reached a certain threshold, it participates in many other processes (biosynthesis of color, aroma, and flavor compounds, specific modification of the cell walls (Tassoni et al., 2006). In RR fruit infection at 3 dpi *LeSAMS3* is suppressed, possibly to pursue a balance of high levels of ET generation that promote fungal colonization. Furthermore, in vegetative tissues, SAM production has been implicated in stress responses like high salinity conditions and ABA treatment (Espartero et al. 1994) and it may play a similar role in fruit responses to pathogen attack.

In the final step of the ET biosynthetic pathway, *LeACO*5 encodes for the enzyme that catalyzes the final conversion of ACC to ET. The expression profile of *LeACO5* in ripening fruit shown by van de Poel et al., (2012), implicated this ACO isoform in the ripening process, as its expression peaked right after the MG stage and is gradually reduced as fruit undergo ripening. An increase in *LeACO*5 during healthy ripening was also observed in our expression analyses and even though it was up-regulated at both time points, the expression is reduced from 1 to 3 dpi. The synthesis of ET appears to be suppressed by a regulatory mechanism at the level of ACO1 transcript and ACO protein abundance. According to van de Poel et al., (2012), increased ACS activity and reduced ACO activity, results in an accumulation of ACC, suggesting that ACO modulates the rate of post-climacteric ET biosynthesis. In MG infection at 1dpi *LeACO*5 presents a significant down-regulation that can be interpreted as a defense strategy to control the increase in ET production. The increase in transcript abundance shown in RR infection may be due to *B. cinerea* itself, as this pathogen has been described to induce ET synthesis in the host, promoting premature senescence or ripening (Marcos et al., 2005; Swartzberg et al., 2008; Cantu et al., 2009).

Ethylene Perception in Tomato Fruit

In general, it is believed that in tomato the effect of ET on defense signaling is ambiguous because it induces a variety of immune responses in conjunction with other signaling networks, but it also promotes senescence or ripening processes, which facilitate infection by pathogens (Cantu et al., 2009; van der Ent and Pieterse, 2012). The ET inhibitor, 1-methylcyclopropene (1-MCP), disrupts ET responses by irreversibly binding to ET receptors and maintaining their phosphorylation state (Kamiyoshihara et al., 2012), therefore it has been widely used to study ripening and disease development in fruit (Blankenship and Dole, 2003; Watkins, 2006; Cantu et al., 2009). Blanco-Ulate et al., (2013) described how ET responses during tomato fruit infection might depend on the concentration and perception of this hormone (Figure 2). In this experiment, tomato fruit at unripe (Mature Green, MG) and ripe (Red Ripe, RR) stages were pre-treated with ET and 1-MCP prior to inoculation with *B. cinerea*.

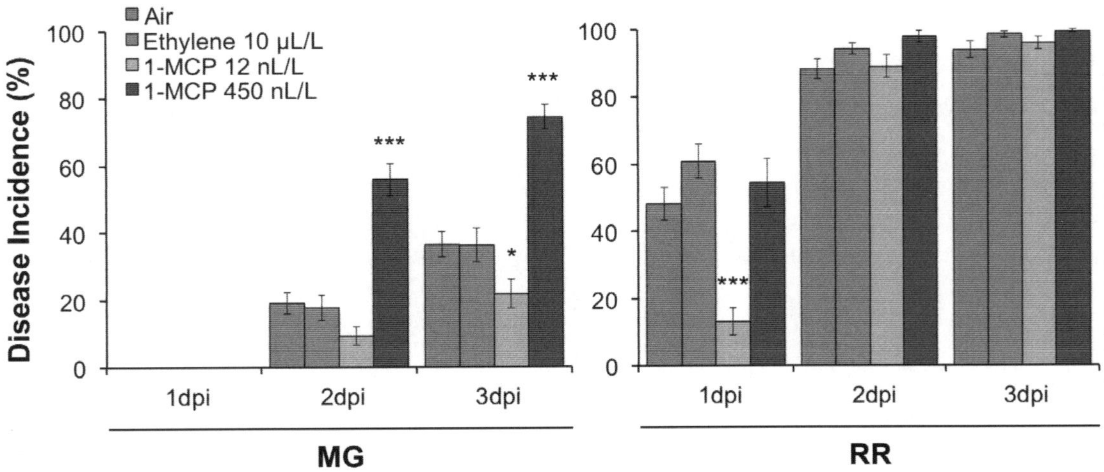

Figure 2. Differences in fruit susceptibility after treatment with low and high concentrations of 1-MCP (Blanco-Ulate et al., 2013)

Pre-treatment of fruit with ET had no effect on infections at either stage. Pre-treatment with low levels of 1-MCP, however, initially reduced disease incidence in both MG and RR fruit. Treatment with high levels of 1-MCP prematurely induced susceptibility in MG fruit but did not influence RR fruit infections. These results suggest that low concentrations of 1-MCP are able to block some ET receptors but not all of them (likely due to limited amounts of inhibitor and continuous generation of receptors). This change in ET perception may promote resistance. High 1-MCP levels, in contrast, possibly block in more effectively most of ET perception, for longer periods of time, if concentrations are high enough, it may also block *de novo* synthetized receptors and consequently hamper resistance response mechanisms that rely on ET perception (Blanco-Ulate et al., 2013).

SA in pathogen-attack: discovery of function through the use NahG fruit

The function of SA in plants was further characterized with the use of *NahG* transgenic plants. They constitutively transgenically express the bacterial *NahG* gene, encoding a salicylic hydrolase, which converts SA into inactive catechol. Our research group, demonstrated that *NahG* expressing transgenic unripe tomato fruit when inoculated with *B. cinerea* (figure 3) display enhanced susceptiblity to infection and did not generate the classical localized necrotic response surrounding the inoculation site that is common in resistant unripe fruit (i.e., a lignified and suberized layer of necrotized cells described in Cantu et al. (2009).

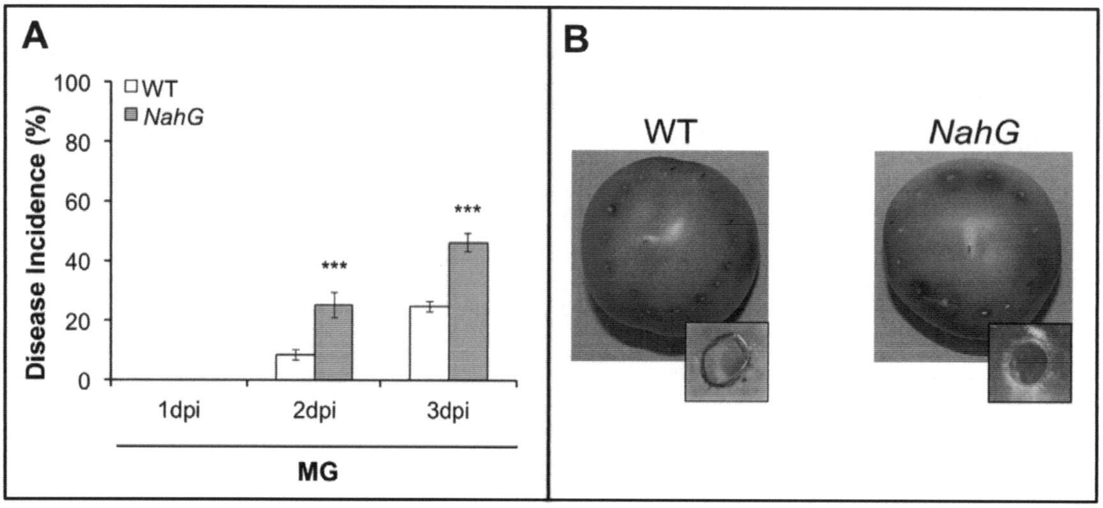

Figure 3. Increase of MG fruit susceptibility observed in SA deficient line *NahG.* A. Disease incidence (%) in MG at 1, 2 and 3 dpi. B. Wild-type and *NahG* MG fruit infected with *B. cinerea* (Blanco-Ulate et al., 2013)

ABA in pathogen-attack: discovery of function through the use of sitiens fruit

ABA-deficient or -insensitive mutants are more sensitive to *P. irregulare* and *Alternaria brassicicola* than wild-type plants. Several groups have reported mutants or transgenic plants with altered resistance to both pathogens and ABA-dependent abiotic stresses (Mengiste et al., 2003; Chini et al., 2004). Moreover, treatment with ABA protects *Arabidopsis* plants against *A. brassicicola* and *Plectosphaerella cucumerina* (Ton and Mauch-Mani, 2004), further implicating a role for ABA in enhanced defense responses against several necrotrophic pathogens. Susceptibility assays with *sitiens* mutant fruit revealed decreased susceptibility in the RR stage. The disease incidence was observed (figure 4A) at 1, 2 and 3 dpi. At the first time point, no significant difference was observed, however, inoculation of RR *sitiens* fruit with *B. cinerea* showed a significant decrease in disease incidence at 2 and 3 dpi when compared to its infected wild type counterpart. This decrease in susceptibility observed by in RR tomato tissue, which is usually colonized by the fungi, suggest that a suppression of ABA synthesis may relate to resistance in late ripening. Additionally, the necrotized tissue that surrounds the wounded site, commonly observed by Cantu et al., (2009) in resistant MG fruit, supports the correlation between ABA and susceptibility in ripe stages of tomato fruit.

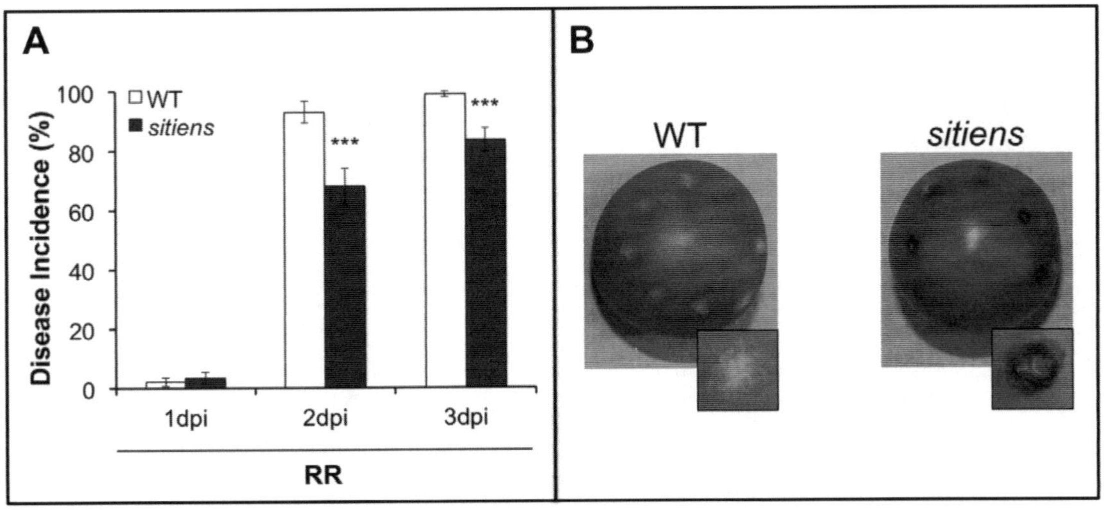

Figure 4. Decreased susceptibility observed in RR fruit from ABA deficient line *sitiens*. A. Disease incidence (%) in RR at 1, 2 and 3 dpi. B. Wild-type and *sitiens* RR fruit infected with *B. cinerea* (Blanco-Ulate et al., 2013)

CONCLUSIONS AND FUTURE STEPS

Extensive transcriptional reprograming of multiple plant hormone networks occurs during the interaction between tomato fruit and *B. cinerea* and during normal uninfected fruit ripening. Our results showed that high levels of ET and ABA, which stimulate senescence and ripening, may facilitate disease development. These processes that can be triggered by pathogenic analogs or induction of its synthesis in the host tissue may play a role in fungal spread. The balance between SA and JA responses seems to be crucial for resistance in MG fruit, while ABA production is associated with the susceptibility of RR fruit. ET, at an appropriate concentration, may contribute to fruit defense by activating JA/ET responses and possibly by blocking the antagonistic effects of SA on JA signaling. At higher concentrations, ET appears to promote fruit ripening and pathogen susceptibility.

This interplay between SA and JA is thought to prioritize the most effective defense responses against necrotrophs like *B. cinerea*. This fungal pathogen has been shown to secrete an elicitor that activates SA responses, which leads to the inactivation of two JA-response genes implicated in resistance against necrotrophs (Pieterse et al. 2009; Alkan et al. 2011; Oirdi et al. 2011; van der Ent and Pieterse 2012). In addition, hormones and/or its derivatives (MeSA) at appropriate concentrations on competent tissues and properly timed, can contribute to resistance strategies by either activating JA, ET or JA/ET-dependent responses or by blocking the SA antagonistic effect on JA-signal transduction. The role of SA in defense response against necrotrophs might be to suppress SAR and through MeSA and SA modulate ET levels to trigger ET-dependent defenses.

ABA appears to be a component of the signaling network activating plant defenses necessary for resistance against some, but not all, necrotrophic pathogens. ABA enhances defenses through modulation of ET, which in turn can regulate JA biosynthesis. In addition, exogenous treatments of ABA induce the expression of the ripening-associated ET biosynthetic genes, triggering ET production and ripening (Zhang et al. 2009). Different biosynthetic genes and/or alternative regulation mechanisms may exist in fruit for these hormones at various stages of ripening, which may result in the activation of multiply regulated responses.

In order for resistant mechanisms to be effective a synchronized, dynamic and complex interaction takes place between these hormones. The outcome of the infection to this pathogen (i.e., fruit is either resistant or susceptible) depends on the interaction of factors that directly or indirectly influence how hormones influence each other in the signaling network. Thus, different elements play more pronounced roles depending on developmental stage, severity and stage of infection, complementary pathogens' attack mechanisms, abiotic stress factors, other signaling molecules and related known players in susceptibility (i.e., cell wall modifying proteins) determine the conditions that develop in a plant-pathogen interaction. Analysis of these biochemical and spatio-temporal factors can provide a more comprehensive analysis of the multiple layers of regulation by hormone signaling networks. In collaboration with Ph.D candidate Mararena Farcuh from the Blumwald Lab, we are interested in translating our hormone-gene related results to plum fruit. Alas, we can get closer to applications that promote resistance during fruit ripening by altering plant hormone signaling pathways.

REFERENCES

Abeles, F. B., P. W. Morgan, and M. E. Saltveit, Jr. (1992). Ethylene in plant biology, 2nd ed. *Academic Press*, San Diego, Calif.

Bagni N. (1989) .Polyamines in plant growth and development. In: Bachrach U, Heimer YM, editors. The physiology of polyamines. Vol. II. Boca Raton, FL: CRC Press; p. 107-120

Bari, R., & Jones, J. (2009). Role of Plant Hormones in Plant Defence Responses. *Plant Mol. Biol.* 69, 473-488.

Bitrián, M., Zarza, X., Altabella, T., Tiburcio, A.F., and Alcázar, R. (2012). Polyamines under Abiotic Stress: Metabolic Crossroads and Hormonal Crosstalks in Plants. *Metabolites* 2, 516-528.

Blanco-Ulate, B., Vincenti, E., Powell, A. L. T., & Cantu, D. (2013). Tomato transcriptome and mutant analyses suggest a role for plant stress hormones in the interaction between fruit and *Botrytis cinerea*. *Frontiers in Plant Science, 4.*

Blankenship, S.M., and Dole, J.M. (2003). 1-Methylcyclopropene: A Review. *Postharvest Biol. Tec.* 28, 1-25.

Boller, T. (1991). Ethylene in pathogenesis and disease resistance, p. 293–324. *In* A. K. Mattoo and J. C. Suttle (ed.), The plant hormone ethylene. *CRC Press*, Boca Raton, Fla.

Cantu, D., Blanco-Ulate, B., Yang, L., Labavitch, J.M., Bennett, A.B., and Powell, A.L.T. (2009). Ripening-Regulated Susceptibility of Tomato Fruit to *Botrytis cinerea* Requires NOR but not RIN or Ethylene. *Plant Physiol.* 150, 1434-1449.

Chini, A., Grant, J.J., Seki, M., Shinozaki, K., and Loake, G.J. (2004). Drought tolerance established by enhanced expression of the CC-NBS-LRR gene, ADR1, requires salicylic acid, EDS1 and ABI1. *Plant J.* 38: 810–822.

Cristescu, S.M., De Martinis, D., Te Lintel Hekkert, S., Parker, D.H., and Harren, F.J.M. (2002). Ethylene Production by *Botrytis cinerea* in vitro and in Tomatoes. *Appl. Environ. Microb.* 68, 5342-5350.

Crozier, A., Y. Kamiya, G. Bishop, and T. Yokota. (2000). "Biosynthesis of Hormones and Elicitor Molecules" in *Biochemistry and Molecular Biology of Plants,* eds. B.B Buchanan, W. Gruissem, and R.L. Jones. Rockville, MD: American Society of Plant Physiologists, 850-929.

Cutler, S. R., Rodriguez, P. L., Finkelstein, R. R., and Abrams, S. R. (2010). Abscisic Acid: Emergence of a Core Signaling Network. *Annu. Rev. Plant Biol. 61*, 651-679.

Davies, P. 2004. Plant hormones: biosynthesis, signal transduction, action! Kluwer Academic Publishers. 3rd ed. 802 p.

Diaz, J., A. ten Have, and J. A. L. van Kan. (2002). The role of ethylene and wounding signaling in resistance of tomato to *Botrytis cinerea. Plant Physiol.* 129:1341–1351.

Espartero, J., Pintor-Toro, J.A., and Pardo, J.M. (1994). Differential Accumulation Of S-adenosylmethionine Synthetase Transcripts in Response to Salt Stress. *Plant Mol. Biol.* 25, 217-227.

Fan XT, Mattheis JP, Roberts RG. (2000). Biosynthesis of phytoalexin in carrot root requires ethylene action. *Physiol Plant* 110:450–454

Hooykaas, Paul JJ, Hall, Michael A, & Libbenga, Kees R. (1999). Biochemistry and molecular biology of plant hormones (Vol. 33): Elsevier Science.

Kamiyoshihara, Y., Tieman, D.M., Huber, D.J., and Klee, H.J. (2012). Ligand-Induced Alterations in the Phosphorylation State of Ethylene Receptors in Tomato Fruit. *Plant Physiol.* 160, 488-497.

Marcos, J.F., González-Candelas, L., and Zacarías, L. (2005). Involvement of Ethylene Biosynthesis and Perception in the Susceptibility of Citrus Fruits to *Penicillium digitatum* Infection and the Accumulation of Defence-related mRNAs. *J. Exp. Bot.* 56, 2183-2193.

Mehta, R.A., Cassol, T., Li, N., Ali, N., Handa, A.K., and Mattoo, A.K. (2002). Engineered Polyamine Accumulation in Tomato Enhances Phytonutrient Content, Juice Quality, and Vine Life. *Nat. Biotech.* 20, 613-618.

Mengiste, T., Chen, X., Salmeron, J., and Dietrich, R. (2003). The BOTRYTIS SUSCEPTIBLE1 gene encodes an R2R3MYB transcription factor protein that is required for biotic and abiotic stress responses in *Arabidopsis*. *Plant Cell* 15: 2551–2565.

Nambeesan, S., Datsenka, T., Ferruzzi, M.G., Malladi, A., Mattoo, A.K., and Handa, A.K. (2010). Overexpression of Yeast Spermidine Synthase Impacts Ripening, Senescence and Decay Symptoms in Tomato. *Plant J.* 63, 836-847.

Pieterse, C. M., Leon-Reyes, A., Van der Ent, S., & Van Wees, S. C. (2009). Networking by small-molecule hormones in plant immunity. Nat Chem Biol, 5(5), 308-316.

Pieterse, C.M.J., Van der Does, D., Zamioudis, C., Leon-Reyes, A., and Van Wees, S.C.M. (2012). Hormonal Modulation of Plant Immunity. *Annu, Rev. Cell. Dev. Bi.* 28, 489 -521

Polevaya, Y., S. Alkalai-Tuvia, A. Copel, and E. Fallik. 2002. Early detection of grey mould development in tomato after harvest. *Postharvest Biol. Technol.* 25:221–225.

Rivas-San Vicente, M., and Plasencia, J. (2011). Salicylic Acid Beyond Defence: Its Role in Plant Growth and Developmen. *J. Exp. Bot.* 62, 3321-3338.

Rodrigo I, Vera P, Tornero P, Hernandez Yago J, Conejero V. (1993). cDNA cloning of viroid-induced tomato pathogenesis-related protein P23: characterization as a vacuolar antifungal factor. *Plant Physiol* 102:939–945.

Swartzberg, D., Kirshner, B., Rav-David, D., Elad, Y., and Granot, D. (2008). *Botrytis cinerea* Induces Senescence and is Inhibited by Autoregulated Expression of the *IPT* gene. *Eur. J. Plant Pathol.* 120, 289-297.

Tassoni, A., Watkins, C. B., & Davies, P. J. (2006). Inhibition of the ethylene response by 1-MCP in tomato suggests that polyamines are not involved in delaying ripening, but may moderate the rate of ripening or over-ripening. J Exp Bot, 57(12), 3313-3325.

Ton, J., and Mauch-Mani, B. (2004). Beta-amino-butyric acid-induced resistance against necrotrophic pathogens is based on ABA-dependent priming for callose. *Plant J.* 38: 119–130.

Van De Poel, B., Bulens, I., Markoula, A., Hertog, M.L. Dreesen, R., Wirtz, M., Vandoninck, S., Oppermann, Y., Keulemans, J., Hell, R., Waelkens, E., De Proft, M.P., Sauter, M., Nicolai, B.M., and Geeraerd, A.H. (2012a). Targeted Systems Biology Profiling of Tomato Fruit Reveals Coordination of the Yang Cycle and a Distinct Regulation of Ethylene Biosynthesis during Postclimacteric Ripening. *Plant Physiol.* 160, 1498-1514.

van der Ent, Sjoerd, & Pieterse, Corné M. J. (2012). Ethylene: Multi-Tasker in Plant–Attacker Interactions Annual Plant Reviews Volume 44 (pp. 343-377): Wiley-Blackwell.

van Kan, J. A. (2006). Licensed to kill: the lifestyle of a necrotrophic plant pathogen. Trends Plant Sci, 11(5), 247-253.

Vandenbussche, F., and Van Der Straeten, D. (2012). "The Role of Ethylene in Plant Growth and Development" in *Annual Plant Reviews* ed. M.T. Mcmanus. Wiley-Blackwell, 219-242.

Wasternack, C. (2007). Jasmonates: An Update on Biosynthesis, Signal Transduction and Action in Plant Stress Response, Growth and Development. *Ann. Bot.* 100, 681-697.

Watkins, C.B. (2006). The Use of 1-Methylcyclopropene (1-MCP) on Fruits and Vegetables. *Biotechnol. Adv.* 24, 389-409.

FIVE YEARS OF STRAWBERRY TRIALS IN COMMERCIAL FIELDS DEMONSTRATE THAT AN EXTRACT OF THE BROWN SEAWEED ASCOPHYLLUM NODOSUM IMPROVES YIELD OF STRAWBERRIES

David Holden[1] and Robin Ross[2]

[1] Holden Research and Consulting, PO Box 1437, Camarillo, CA, USA

[2] Acadian Seaplants Limited, 30 Brown Avenue, Dartmouth, NS, Canada

Strawberry growers are often looking for sustainable products to enhance the overall health and yield of their crop. Over five years, one summer and four winter studies, conducted in the Oxnard area of California, showed that soil drip-tape applications of seaweed extract (Acadian) enhanced yield of strawberries by an average of 395 trays per acre compared to the control. This improvement in yield are likely due to improvements in early growth and crown development as were seen in the first three years of studies. There was also suppression of mites in the final two years of trials. Charcoal Rot caused by the pathogen *Macrophomina paseolina* was present in the fourth year of this study (2013-2014). The seaweed extract treated plots had less mortality due to this disease compared to the control (27.4 vs 16.6 percent plant mortality). Lastly, there were high soil sodium levels in the first year study and the seaweed extract treated plots had less phytotoxicity as well as lower sodium levels in the leaves. These results indicated that seaweed extract significantly improves yield of strawberries however the mode of action is likely multifaceted and may involve improvements in growth, tolerance to environmental stresses, and tolerance to biotic stresses.

Total Marketable Yield

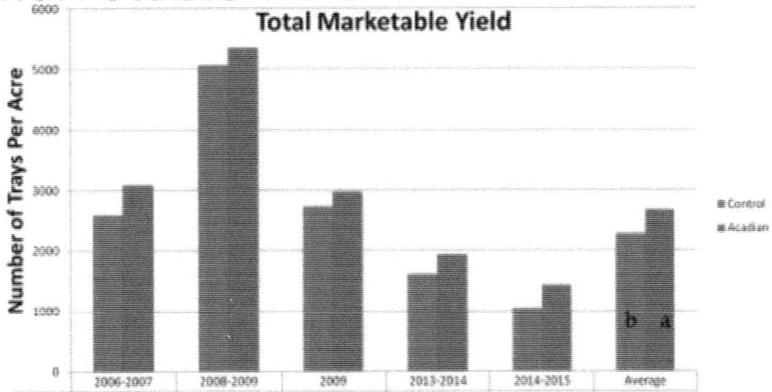

	2006-2007	2008-2009	2009	2013-2014	2014-2015	Average
Control	2598.2	5066.6	2733.7	1618.0	1043.0	2273.0
Acadian	3087.3	5348.9	2969.6	1932.0	1426.0	2668.0

Total yield was enhanced all four years. The average yield increase was 395 trays per acre which is a 17% yield increase.
Means followed by the same letter are not different, Duncan's new MRT, p=.02

Numbers of Crowns Per Plant

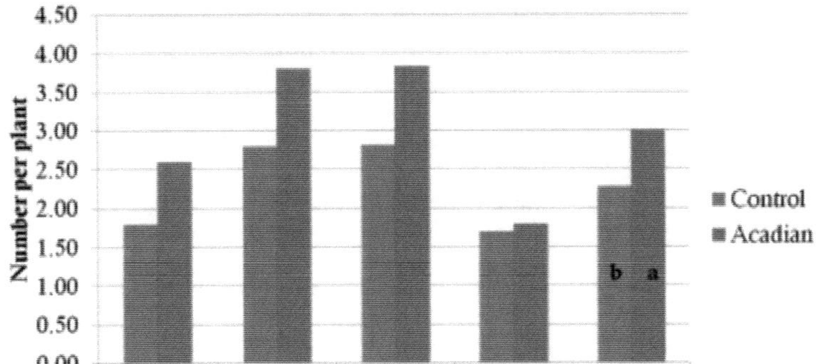

	2006-2007	2008-2009	2009	2014-2015	Average
Control	1.80	2.80	2.81	1.70	2.28
Acadian	2.60	3.80	3.84	1.80	3.01

Means followed by the same letter are not different, Duncan's New mrt, p=.04
There were significantly more crown divisions by the beginning of the harvest season (4 months after planting)

2013-2014 Trial
Macrophomena phaseolina, Charcoal Rot
Percent Plant Mortality

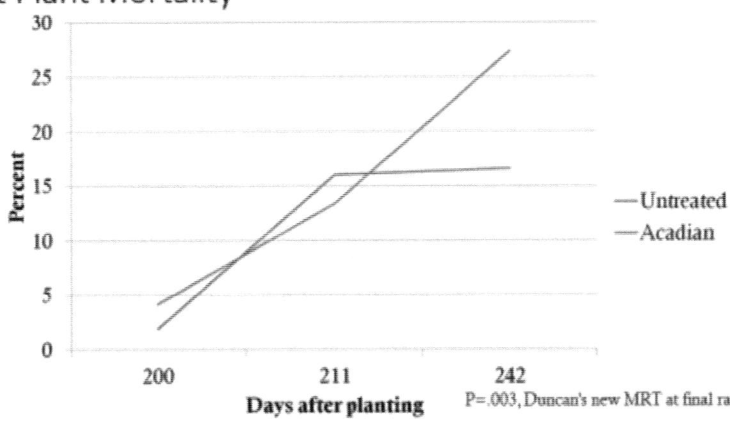

P=.003, Duncan's new MRT at final rating.

FRUIT THINNING: A LONG PROCESS TO FIND REPEATABLE AND ACCEPTABLE METHODS IN ORDER TO THIN FRUIT

Dr. Karl-Friedrich Bangerth, University of Hohenheim

Thinning of fruit has been practised for more than 3000 years (Dennis, 2000). Efforts to find acceptable and practical methods to thin fruit in a reliable way date back more than a century (for a good review see Addicott, 1982 and Plant Growth Regulation 2000, Volume 32). Although considerable progress has been made from 1982 until today, as far as reliability and the production of good fruit size, quality and yield are concerned, fruit thinning remains a formidable challenge for orchardists to achieve an optimal or the most desired outcome. The historical perspectives presentation will concentrate mainly on apple fruit thinning.

FIELD EFFICACY OF AMINOETHOXYVINYLGLYCINE (RETAIN®) FOR IMPROVING FRUIT SET OF ALMOND

Robert Fritts Jr[1], Mike Ansolabehere[2], and Tino Lopez[2]

[1]Valent BioSciences Corporation, 870 Technology Way, Libertyville, IL 60048, USA

[2]Valent USA Corporation, 1600 Riviera Ave, Suite 200, Walnut Creek, CA 94596, USA

SUMMARY

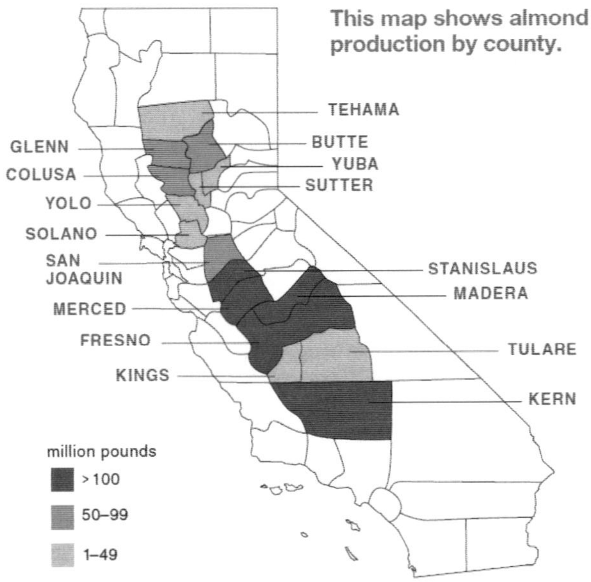

Field trials were conducted in California during 2014 to evaluate AVG/ReTain® for increase of fruit set and yield of almonds. Based on fruit per limb cross sectional area, percent fruit set counts, and harvest weights, ReTain® treatment increased almond fruit set and yield at all test sites in California.

OBJECTIVE

To evaluate the large scale field efficacy of ReTain® for increased fruit set of almond in California.

INTRODUCTION

California is the major almond producing state in the United States and produces about 82% of the world's almonds and 100% of the U.S. supply. Almonds are grown as far north as Tehama County and as far south as Kern County.

California's 2013 almond acreage was estimated at 940,000 acres, up 1 percent from the 2012 acreage of 930,000. Of the total acreage for 2013, 840,000 acres were bearing and 100,000 acres were non-bearing. Preliminary bearing acreage for 2014 is estimated at 860,000. Nonpareil continues to be the leading variety at 38 percent, followed by Butte, Monterey, Carmel and Padre. Kern, Fresno, Stanislaus, Merced and Madera were the leading counties. These five counties had 74 percent of the total bearing acreage. (CDFA, 2014).

Almonds bloom from early February through March depending on location. For almond to produce a crop the tree must be cross pollinated with a different variety of almond. Growers will plant multiple varieties within the same orchard to help increase the potential of setting a larger crop. Honey bees are also placed in the orchard well before the first blooms to help in pollination and fruit set. The almond blossom is only viable for a short period of time, less than one week. If the weather is cool or wet, or the bloom is not synchronized with a cross pollinating variety, fruit set is reduced dramatically.

New plantings are being made each year to cope with the world demand for the crop. Growers are always seeking ways to increase yields and ReTain® is a tool that can help them in that regards. Another factor, is that by increasing yield on existing plantings there could be a reduction of new plantings. This could help the current drought condition that exist in CA and also reduce other inputs such as pesticides and fertilizers on a state-wide basis.

Aminoethoxyvinylglycine (aviglycine, AVG) is a potent inhibitor of 1-aminocyclopropane-1-carboxylic acid (ACC) synthase and thus effectively blocks ethylene synthesis (Yu et al., 1979). AVG has been commercialized by Valent BioSciences Corporation (Libertyville, Illinois) as ReTain® and is commonly used to delay ripening and prevent fruit drop of apple (Silverman 2004), reduce pistillate flower abscission of walnut (Retamales and Petracek, 2010), and prevent natural induction of inflorescence development of pineapple (Bartholomew et al., 2011).

Recent research at Ohio State University (Racsko, 2012) has shown that AVG/ReTain® can maintain the viability of cherry blooms for an additional three to four days. In these studies, AVG treatments delayed whole flower and stigmatic senescence evidenced by a delay in flowering phenological stages of both detached and attached flowers, and by scanning electron microscopy imaging of the stigmatic surface. AVG/ReTain® treatments did not affect floral biological characteristics (intra-floral nectar production, nectar refraction, pollen production, and pollen viability) of cherry flowers, and therefore did not influence the attractiveness of flowers to insect pollinators. Ethylene evolution of detached cherry flowers and fruit drop were significantly decreased. While fruit set was significantly increased, significant differences in fruit quality (fruit size, color, soluble solids concentration) were not observed between control and AVG/ReTain® treatments (Racsko, 2012). This increased flower viability has been shown to allow for improved fruit set of cherry flowers.

Sweet cherry (*Prunus avium*) and almond (*Prunus dulcis*) are members of the same genera. It was, therefore, appropriate to determine if their floral parts evolved ethylene in a similar manner and, concomitantly, responded the same to AVG/ReTain® applications. Multiple large-scale field trials were initiated in major almond growing regions in California to evaluate AVG/ReTain® promotion of almond fruit set. This report provides a summary of results obtained in almond during the 2014 season.

MATERIALS AND METHODS

Eight commercial almond orchards were selected for field trials (**Table 1**), with several of the sites having multiple varieties. Varieties tested were Nonpareil, Monterey, Aldridge and Fritz. Nonpareil, the most common almond variety, represented the majority of the trials conducted.

Table 1. Location to the nearest town, county, and variety of almond tested for field trials with ReTain® for improved fruit set during 2014 in California.

Nearest Town	County	Variety
Tulare	Tulare	Nonpareil, Monterey
Easton	Fresno	Nonpareil, Monterey
Madera	Madera	Nonpareil, Fritz, Monterey
Ripperdan (30)	Madera	Nonpareil
Ripperdan (145)	Madera	Nonpareil
Chowchilla	Madera	Nonpareil
Kerman	Fresno	Nonpareil, Aldridge
Terra Bella	Tulare	Nonpareil

Application timing was based on the previous research report by Racsko (2012), and experience with the same use pattern for ReTain® Plant Growth Regulator for California on walnuts and cherry, which targets the blossoms when they are in the initial stages of bloom, i.e., popcorn to early bloom. On a whole tree basis, this translated to approximately 15 to 20 percent open blossoms (**Figure 1**).

Figure 1. Bloom stage for application timing of ReTain® for increased fruit set of almond.

A five-acre block within the orchard was selected for each trial. A single application of ReTain® alone at one pouch (50 g active ingredient, AI) per acre was applied using a commercial air blast sprayer at one hundred gallons of water per acre. As most commercial sprayers have a volume of 500 gallons, the treated area was conveniently equivalent to one spray tank. This method assured that the appropriate water volume and product per acre was consistent from location to location. An adjacent area of untreated control (UTC) trees was selected for comparison.

At three of the locations (encompassing four sites and two varieties), branches were tagged with ribbon and the number of blooms were recorded. Once fruit was well established, the same branches were again counted to determine the percent fruit set. Data was analyzed using each of the trial sites as a replicate and comparing the ReTain® treatment to non-treated trees.

At five sites (encompassing 9 trials) fruit set was evaluated once fruit was well established. Fruit set measurements consisted of selecting a single branch on a tree that radiated out from the central trunk of the tree. Branch sizes were from 0.2 centimeters to 0.9 centimeters in diameter. All the fruit from the base of the limb to the end on the selected limb were counted. The circumference of each evaluation limb was measured with a tape measure calibrated to convert the measurement into a diameter value. At each site, single limbs of 40 separate trees were evaluated per treatment. Trees were selected at random to represent the length and width of the respective plot. The data was entered into an Excel spreadsheet and limb cross sectional area was calculated for each individual limb. The amount of fruit set was then determined by dividing the number of fruit on an individual limb by the corresponding limb cross sectional area. The resultant number is expressed as fruit per centimeter limb cross sectional area (fruit/LCSA). The data is presented in **Table 3**.

At three locations, the effect of varying ReTain® rates on fruit set of almonds were tested. Use rates of 50, 75, or 100 g AI per acre of ReTain® were applied using a commercial air blast sprayer at one hundred gallons of water per acre. An adjacent area of untreated trees was selected for comparison. At each site, single limbs of 40 separate trees were evaluated per treatment. Trees were selected at random to represent the length and width of the respective plot. Fruit set measurements and data collection were determined as previously described.

Almond yields were evaluated in seven commercial almond orchards. The trees were treated with bloom applications of ReTain at a rate of 333 grams of formulated product per acre, using commercial spray equipment calibrated to deliver 100 GPA. Blocks were four to five acres in size. The total harvestable nuts for the treated and untreated blocks were collected and weighed using commercial harvesting machinery.

A two-sample t-test was used to test for statistical differences in fruit set between ReTain® treated and non-treated treatment blocks. An additional ANOVA test was performed on the pooled data using each of the trial sites as a replicate and comparing the ReTain® treatment to non-treated trees for the average of all trial sites.

RESULTS

There was a significant increase in fruit set from ReTain® treatment at all the trial sites and varieties. For percent fruit set (**Table 2**), ReTain® significantly increased fruit set by 33.3 percent over the four sites. The difference in crop load as a percentage increase in fruit per cross sectional area between ReTain® treated trees and non-treated trees shows a range from a low of 18.1% to a high of 55.4% with an average increase of 28.4% over all nine trial sites (**Table 3**).

Table 2. Percent fruit set from counts taken on tagged shoots at bloom and following fruit establishment in almond following a 50 g AI ReTain® application.

Percent Fruit Set in Almond Following Bloom Application of ReTain®				
Location	Almond Variety	Percent Fruit Set		Percent Increase
		Untreated Control	ReTain® at 50 g AI/A	
Chowchilla	Nonpareil	29.3	35.3	20.5
Terra Bella	Nonpareil	21.0	32.3	53.8
Kerman	Aldridge	27.6	40.9	48.2
Kerman	Nonpareil	38.5	46.7	21.3
Average of all sites[y]		29.1 a	38.8 b	33.3

[y] Data from all sites were analyzed using a two-sample t-test for statistical differences at the 5% level.

Table 3. Effect of a 50 grams AI ReTain® treatment on almond fruit set, expressed as fruit per limb cross sectional area (fruit/LCSA) and percent increase, compared to non-treated trees in California during 2014.

Percent Fruit Set in Almond Following Bloom Application of ReTain®					
		Fruit/LCSA (cm^2)			
Location	Variety	Treated	Untreated	Percent increase	Significance[z]
Tulare	Nonpareil	50.6	39.2	29.1	**
	Monterey	64.8	54.7	18.5	**
Madera	Fritz	32.9	27.1	21.4	**
	Nonpareil	32.8	21.1	55.4	**
	Monterey	40.1	28.2	42.0	**
Easton	Monterey	38.3	30.8	24.2	**
	Nonpareil	37.1	31.4	18.1	**
Ripperdan (145)	Nonpareil	44.5	34.8	28.1	**
Ripperdan (30)	Nonpareil	46.0	34.0	35.4	**
Average of all sites[y]		43.0 a	33.5 b	28.4	

[z] A double asterisk designates a 95% confidence that there is a statistical difference between the ReTain® treated and non-treated trees.

[y] Data from all sites was analyzed using Duncan's multiple range test at the 5% significance level.

At three locations, the effect of varying ReTain® rates on fruit set of almonds were tested. A summary of results are presented in **Table 4**.

Table 4. Effect of a 50, 75, and 100 g AI ReTain® on almond fruit set, expressed as fruit per limb cross sectional area (fruit/LCSA) and percent increase, compared to untreated trees in California during 2014.

Location	Variety	Treatment	Fruit / LCSA (cm^2)	Percent Increase Compared to UTC
Ripperdan (30)	Nonpareil (UTC=33.96)	50 g AI	45.99	35.4%
		75 g AI	50.79	49.6%
Ripperdan (145)	Nonpareil (UTC=34.76)	50 g AI	44.52	28.1%
		75 g AI	45.91	32.1%
Easton	Monterey (UTC=30.82)	50 g AI	38.30	24.3%
		100 g AI	44.52	44.5%
Easton	Nonpareil (UTC=31.41)	50 g AI	37.07	18.0%
		100 g AI	35.22	12.1%

Two sites of the Nonpereil variety near Ripperdan, in Madera County, compared 50 and 75 g AI per acre (one pouch, and one and a half pouches, respectively) to untreated trees. The data for these sites is shown in **Figure 2**.

Figure 2. Effect of a 50 and 75 grams AI ReTain® treatment on almond fruit set, expressed as fruit per limb cross sectional area (fruit/LCSA), compared to non-treated trees in Ripperdan, Madera County, California during 2014[z].

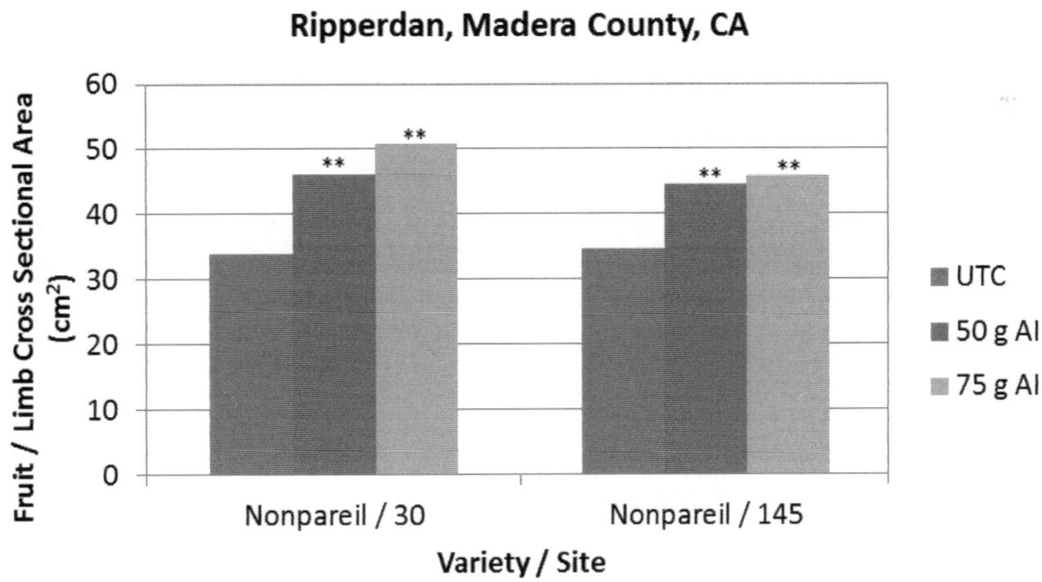

[z] A double asterisk designates a 95% confidence that there is a statistical difference between the ReTain® treated and non-treated trees using a two-sample t-test statistical analysis.

An additional two sites, one Monterey and one Nonpereil variety, near Easton in Fresno County, compared 50 and 100 g AI per acre (one pouch and two pouches, respectively) to untreated trees. The data is shown in **Figure 3**.

Figure 3. Effect of a 50 and 100 grams AI ReTain® treatment on almond fruit set, expressed as fruit per limb cross sectional area (fruit/LCSA), compared to non-treated trees in Easton, Fresno County, California during 2014[z].

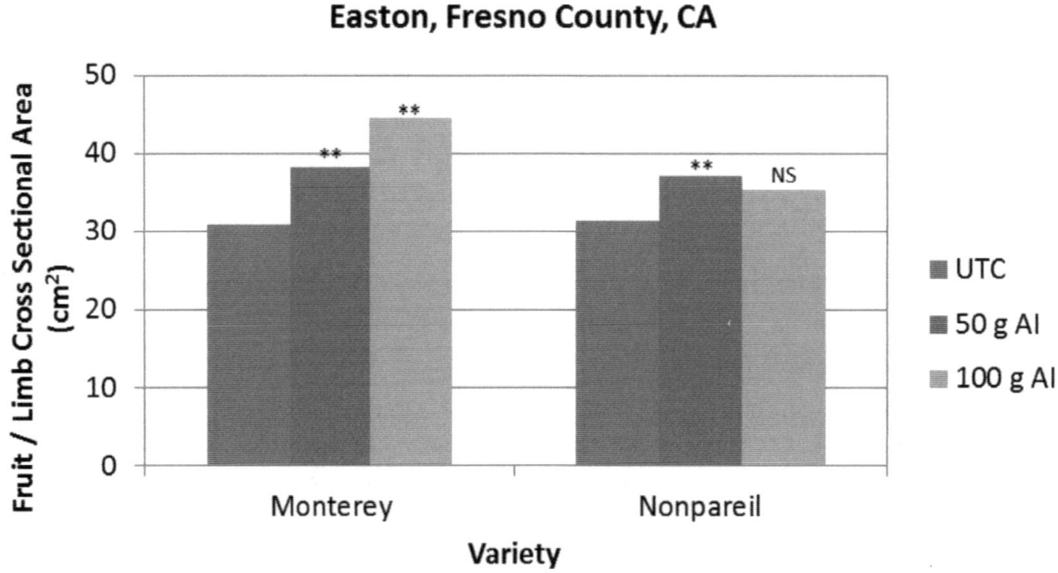

[z] A double asterisk designates a 95% confidence that there is a statistical difference between the ReTain® treated and non-treated trees using a two-sample t-test statistical analysis.

Both of the Nonpareil varieties at Ripperdan, Madera County sites showed an increase in fruit set with an increased rate of ReTain®. At the Fresno County site, only the Monterey variety showed an increase in fruit set with the higher ReTain® use rate. The Nonpareil variety at this site was not significantly better at the 100g AI use rate than the untreated control.

Almond yields were evaluated in seven commercial almond orchards using commercial harvesting equipment. In all cases there was an increase in harvested nuts, however, not always statistically significant. When the seven sites were pooled and each site evaluated as a replicate, the average of all sites showed a statistically significant increase in yield of 217 pounds for Nonpareil and 305 pounds for the pollinator varieties with ReTain® application at bloom (**Table 5 and 6**).

Table 5. Yield of Nonpareil almonds showing trial location, tree age, and percent bloom at the time of application for field trials treated with ReTain at a rate of 333 grams of formulated product per acre.

Location	**Tree Age (years)**	**% Bloom at App.**	**Pound Nut Meat/Acre**			**Significance**
			Untreated	*ReTain*	Difference	
Terra Bella	20	5	2771	3150	+ 379	5%
Chowchilla	47	20	2011	2224	+213	5%
Easton	4	20	1614	1885	+271	not replicated
Tulare	5	10	1195	1389	+194	NS
Shafter	9	5	3065	3265	+174	NS
Kerman	10	10	3494	3561	+67	NS
Madera Ranchos	10	15	2731	2926	+195	NS
Averages of all trials [y]			2412 b	2629 a	+217	1%

[y] Data from all sites was analyzed using Duncan's multiple range test at the 5% significance level.

Table 6. Yield of pollinator varieties of almonds showing trial location and percent bloom at the time of application for field trials treated with ReTain at a rate of 333 grams of formulated product per acre.

Variety (No. of sites)	Pound Nut Meat/Acre			Significance
	Untreated	*ReTain*	Difference	
Monterey (4)	2476	2772	+296	10%
Fritz (3)	2476	2772	+296	10%
Aldrich (1)	1983	2757	+774	5%
Average all sites [y]	2539	2844	+305	5%

No phytotoxicity to almond trees was noted in any of the trials.

CONCLUSION

Application of ReTain® during bloom significantly increased fruit set of almond trees in California in the thirteen sites evaluated. The data was highly significant when viewed as a comparison of treatments using fruit per limb cross sectional area and percent fruit set within a given site, or pooling the data across all sites. In addition, a moderate rate effect was evident in the three sites where multiple rates were tested.

There are many factors that influence the final yield of an almond crop. The major factors being crop load, nutrition, and irrigation quantity and quality. The field trials presented in this paper demonstrate that even without additional inputs, ReTain® application increased fruit set of almonds, which translated into an increase in yield at harvest time.

REFERENCES

Bartholomew, D. P., G. Uruu, J. A. Lopez and D. Leep, 2011. Refining aviglycine treatments to improve control of natural induction of 'MD-2' pineapple. Acta Hort. 90-2: 211-220.

California Department of Food and Agriculture (CDFA), USDA, NASS, Pacific Region. 2013 California Almond Acreage Report, April 24, 2014

Racsko, J. 2012. Improving cherry fruit set in 'Regina' sweet cherry with aminoethoxyvynilglycine (ReTain®) is facilitated by delayed stigmatic senescence. HortScience 47(9): S155.

Retamales, J. and P. D. Petracek, 2010. AVG and fruit set: a tool for which novel applications are still being developed in various fruit crops - the case of walnuts. Acta Hort. 884: 337-341.

Silverman, F. P., P. D. Petracek, M. R. Noll and P. Warrior, 2004. Aminoethoxyvinylglycine effects on late-season apple fruit maturation. Plant Growth Reg. 43: 153-161.

Yu, Y.-B., D. O. Adams and S. F. Yang, 1979. 1-Aminocyclopropanecarboxylate synthase, a key enzyme in ethylene biosynthesis. Arch. Biochem. Biophys. 198: 280–286.

TRUNK INJECTION OF SOME NOVEL CHEMICAL THINNING AGENTS INTO APPLE TREES

Steven McArtney, Jozsef Racsko, Valent BioSciences Corporation, 870 Technology Way, Libertyville, IL, USA

J.D. Obermiller, North Carolina State University, 455 Research Drive, Mills River, NC, USA

Duane Greene, University of Massachusetts, 80 Campus Center Way, Amherst, MA, USA

Sam Drahn, Rainbow Treecare Scientific Advancements, 11571 K-Tel Drive, Minnetonka, MN, USA

Airblast spray application technology is an effective but inefficient method to deliver thinning chemicals in apple orchards. Spray interception values from airblast sprayers is typically less than 40 percent at the time when thinners are applied. Even after spray droplets land on the target, penetration of polar active compounds through the lipophilic plant cuticle can be 20 percent or less. Trunk injection was investigated as a method for delivering systemic thinning chemicals to apple trees that eliminates off-target drift and minimizes the risk of environmental contamination. The ethylene precursor 1-aminocyclopropane-1-carboxylic acid (ACC), the PSII inhibitor metamitron (MET), or S-Abscisic acid (S-ABA) were injected into the trunk of mature 'Cripps Pink'/M.7 apple trees when the mean fruit diameter was 6 mm. Thinning activity of these treatments was compared to foliar airblast spray application at the same time. Ethylene evolution from detached fruit was increased 24 h after trunk injection of ACC, and final fruit set was reduced, however injection was less effective than a foliar ACC spray. Trunk injection of S-ABA reduced both stomatal conductance 24 h after treatment and final fruit set more effectively than a foliar spray. Trunk injection of MET reduced fruit set more effectively than foliar a spray. Significant phytotoxicty was observed following high rates of injection of ACC or MET, whereas foliar spray application of the same materials at an equivalent dosage per tree resulted in minimal phytotoxicity. In addition to a thinning response, trunk injection of S-ABA provided season-long control of vegetative growth. These results describe the potential for trunk injection as an alternative method for delivering fruit abscission agents to perennial trees.

PRELIMINARY EVALUATION OF EFFICACY OF AMINOETHOXYVINYLGLYCINE HYDROCHLORIDE (AVG, RETAIN®) SPRAYS TO INCREASE FRUIT SIZE OR DECREASE MATURE FRUIT DROP IN PURPLE-FRUITED PITANGA (EUGENIA UNIFLORA L.)

John L Griffis, Jr. Florida Gulf Coast University, Fort Myers Florida USA

Ty G. McDonald University of Hawai'i at Manoa, Honolulu Hawai'i USA

Malcolm M. Manners Florida Southern College, Lakeland Florida USA

The pitanga (*Eugenia uniflora* L.) is a tropical fruit native to Brazil, but distributed in the tropics worldwide (Morton & Dowling, Jr., 1987). Although pitanga (red or purple-fruited) can grow and bear fruit in many different locations, it is not commercially produced anywhere except in Brazil, although there are small-scale producers in other locations such as Hawai'i (Griffis *et al.*, 2009). There are many reasons why the crop has not been more widely adopted. Among those would be a relatively large seed-to-flesh ratio (about 20%/80%) and asynchronous blooming, such that unopened buds, open flowers, immature fruits in various stages of development and mature fruits are present on individual plants at the same time. The fruits are non-climacteric and the largest, best-tasting fruits must be harvested when fully ripe; immature fruits are generally unpalatable. Unfortunately the fruits tend to drop quickly from the plants after achieving maturity. Very frequent harvesting is essential to reducing crop loss due to fruit drop (Griffis *et al.*, 2013).

AVG (ReTain®), an ethylene biosynthesis inhibitor, could be utilized either to increase fruit size and flesh-to-seed ratio as it does with some other fruit crops (Byers *et al.*, 2005; Rath and Prentice, 2004), and/or to delay fruit drop and fruit softening as it does in some other fruit crops (Bregoli *et al.*, 2002; Stover *et al.*, 2003) so that harvesting of the purple pitanga fruits could be less frequent and more efficient. However, nothing has been reported concerning the efficacy of AVG sprays on any members of the Myrtaceae so that phytotoxicity, the rates of application, frequency of application and use of spreader-stickers (or not) all needed to be determined. Preliminary trials were conducted in our purple-fruited pitanga plots in Hawai'i and Florida to determine if ReTain® sprays might be effective for controlling fruit drop or increasing fruit size of this crop.

Materials and Methods

The pitanga plants in our fields in Florida and Hawai'i are mostly seedlings, so some variability in the results was expected. Our first trial, conducted at the Kona Experiment Station in Hawai'i in Spring 2015, tested for any phytotoxicity that ReTain® or the spreader-sticker Freeway® might have on purple-fruited pitanga. Three randomly selected seedling plants from within the field of about 130 plants were sprayed with water. Another three plants were sprayed with a ReTain® solution calibrated to provide 250ppm a.i. AVG. Another three plants were sprayed with a solution of water and 0.5ml/L of Freeway® spreader-sticker.

The second trial was conducted in Florida, where our original grafted plant of the variety 'Zill Dark' is located. This original plant is isolated from the other pitanga plantings on the Florida Southern College campus and is often used for budwood to produce grafted plants of this variety. The fruits from this plant are also frequently harvested and stored frozen for further evaluation and comparison against other fruits from seedling plants. In Spring 2014, when this plant was not treated with any plant hormone, several large bags of purple fruits from this original 'Zill Dark' plant were harvested and frozen. One hundred ripe fruits harvested from this plant were weighed and the seed to pulp ratios were determined. In late Spring 2015, when some fruits had already matured, this plant was sprayed with a combination solution of 250 ppm a.i. ReTain® (1.67 g/L) and 0.5 ml/L of Freeway® spreader-sticker. Sixty ripe fruits were harvested one week later. On the day of this harvest, the plant was sprayed again with the same solution and forty ripe fruits were harvested one week later. Another spray of this same solution was applied fourteen days later and one week afterward, 20 ripe fruits were harvested. All fruits were weighted and the seed to pulp ratios were determined. The plant was also evaluated for the number of fruit mummies that remained after all ripe fruits had either dropped or been removed at the end of the season.

The third trial was again conducted in Hawai'i in Spring 2015, where there are also 20 grafted plants of 'Zill Dark' planted adjacent to the seedling field. In this trial, three 'Zill Dark' plants were sprayed with water until run-off on March 3, 2015. Three other 'Zill Dark' plants were sprayed until run-off with a combination solution of 250 ppm a.i. ReTain® and 0.5 ml/L of Freeway® spreader-sticker one time on March 3, 2015. Three other 'Zill Dark' plants were sprayed until run-off with a combination solution of 250 ppm a.i. ReTain® and 0.5 ml/L of Freeway® spreader-sticker once on March 6, 2015, again on March 13, 2015 and a third time on March 20, 2015. Fruits were harvested and counted from all the plants from March 20 to May 29, 2015; harvests were conducted about twenty times for each plant. Average fruits weights were also determined at some of the harvests.

Results and Conclusions

There were no signs of phytotoxicity in the first trial, so we were able to proceed with the next two trials. In the trial conducted in Florida, the one hundred purple fruits harvested from the unsprayed 'Zill Dark" were determined to have an average weight of 5.34g and a seed-to-fruit ratio of 15.6% seed to 84.4 % pulp. The sixty purple fruits harvested seven days after the single spray to run-off of ReTain® and Freeway® were determined to have an average weight of 7.19g, or about a 35% increase in fruit size.

The forty purple fruits harvested fourteen days after the first spray of ReTain® and Freeway® and seven days after the second spray of ReTain® and Freeway® were determined to have an average weight of 5.89g and a ratio of 16.7% seed to 83.3 % fruit pulp. The twenty fruits harvested twenty-eight days after the first spray of ReTain® and Freeway®, twenty one days after the second spray of ReTain® and Freeway® and seven days after the third spray of ReTain® and Freeway® were determined to have an average weight of 5.98g and a ratio of 18% seed to 82% fruit. The number of fruit mummies found on the plant two weeks later was considerably more than is normally found; the plant usually produces very few if any mummies.

The third trial had considerably more fruit over the harvest period that lasted more than two months. The average 'Zill Dark' purple pitanga fruit size in the Hawai'i was greater than it was in Florida. The average weight of several fruits collected from the plants sprayed with water on March 3, 2015 and harvested on April 27, 2015 was 9.47g. The average weight of several fruits sprayed once with the ReTain® and Freeway® solution on March 3, but harvested on April 27, 2015 was 6.6g. The average weight of several fruits sprayed with the ReTain® and Freeway® three times on March 6, March 13, and March 20, 2015 and harvested on April 27, 2015 was 7.9g. The total number of fruits harvested in this trial was 17,189. The three plants sprayed with water only produced 5,336 ripe fruits during the trial period. The three plants sprayed one time with ReTain® and Freeway® solution produced 7,385 fruits during the trial period. The three plants sprayed three times with the ReTain® and Freeway® solution produced 4,468 fruits during the trial period. The average yield (by weight) of the three plants sprayed with just water was 17.1 kg/plant over the trial period. The average yield (by weight) of the three plants sprayed one time with the ReTain® and Freeway® solution was 16.1 kg/plant over the trial period. The average yield (by weight) of the three plants sprayed three times with the ReTain® and Freeway® solution was 11.7k kg/plant over the trial period.

The overall results of the trials are rather mixed and the project needs to be repeated or expanded to include more grafted 'Zill Dark' plants and more fruits need to weighed at all harvest dates to determine how fruit size varies with relation from time of ReTain® and Freeway® spray and harvest date. The variability of fruit size and fruit quantity data for purple fruited pitanga have be documented (de Lira Junior *et al.*, 2010) and more than one year of data will be needed to draw any substantial conclusions of the efficacy of ReTain® (AVG) in increasing fruit size or reducing fruit drop in purple fruited pitanga. A more precise count of fruit mummies that remain after all ripe fruits have been harvested may also allow us to determine if AVG is effective in reducing fruit drop.

Literature Cited

Bregoli, A. M., Scaramagli, S., Costa, G., Sabatini, E., Ziosi, V., Biondi, S., & Torrigiani, P. (2002). Peach (Prunus persica) fruit ripening: aminoethoxyvinylglycine (AVG) and exogenous polyamines affect ethylene emission and flesh firmness. *Physiologia Plantarum, 114*(3), 472-481.

Byers, R. E., Carbaugh, D. H., & Combs, L. D. (2005). Ethylene inhibitors delay fruit drop, maturity, and increase fruit size of'Arlet' apples. *HortScience, 40*(7), 2061-2065.

de Lira Júnior, J. S., Fernandes Bezerra, J. E., & Lederman, I. E. (2010). Repetibilidade da produção, número e peso de frutos de seleções de pitanga roxa. *Acta Agronómica, 59*(1), 103-110.

Griffis, J.L. Jr., McDonald, T.G., Smith, V.E., & Manners, M.M. (2009). *Eugenia uniflora*--a nutritious, easy-to-grow fruit for the tropics. *Acta Horticulturae,* (806):277-284.

Griffis, J. L. Jr., Sams, C.E., Manners, M.M., McDonald, T.G. and Radovich, T.J.K. (2013) A progress report on commercialization in the USA of purple-fruited pitanga (*Eugenia uniflora* L.), an underutilized fruit crop. *Acta Horticulturae*, (979):807-814.

Morton, J.F. & Dowling, Jr., C.F. (1987). *Fruits of warm climates / Julia F. Morton; edited by Curtis F. Dowling, Jr.* J.F. Morton; Distributed by Creative Resources Systems.

Rath, A. C., & Prentice, A. J. (2004). Yield increase and higher flesh firmness of'Arctic Snow'nectarines both at harvest in Australia and after export to Taiwan following pre-harvest application of ReTain Plant Growth Regulator (aminoethoxyvinylglycine, AVG). *Animal Production Science*, *44*(3), 343-351.

Stover, E., Fargione, M. J., Watkins, C. B., & Iungerman, K. A. (2003). Harvest Management of Marshall McIntosh' Apples: Effects of AVG, NAA, Ethephon, and Summer Pruning on Preharvest Drop and Fruit Quality. *HortScience*, *38*(6), 1093-1099.

PLANT GROWTH REGULATION NEEDS IN ORNAMENTAL NURSERY PRODUCTION

Dave Barcel

OHP Inc., Genesee, WI

Plant Growth Regulators (PGR'S) are used commonly on ornamental crops such as bedding plants, flowering potted plants and perennial crops. PGR use on woody ornamental crop such as trees or shrubs was typically not performed. Many nursery crops tend to be long-term crops (1 to 3 years) and as such, growers need to finish production as soon as possible for sale.

The downturn of the nursery market in 2009 and subsequent years forced growers to inventory plant material longer. Growers then needed to slow down and hold growth on plants do to the poor sales environment. The Gibberellin inhibiting PGR'S such as the triazole group (Bonzi, Paczol, etc.) can provide the growth control needed for these crops. Not all PGR's are labeled for outdoor use, and some PGR'S will have less effectiveness than others. Thus growers must make sure they use the correct product having the correct labeled use for their needs.

Other needs of the nursery business include increased branching of crops such as hydrangea, knockout rose, young tree whips, perennials, etc. Pruning of these crops multiple times during the production season is the traditional way to increase branching. There are a few PGR'S that can induce branching via chemical treatments. These products work by reducing apical dominance, inducing bud or tissue growth or eliminate floral production in favor of vegetative production. Products such as Augeo, Florel and Configure are used in the nursery industry for such uses.

Lastly, there is a need for nursery crops being dug and stored dormant in cold storage to be harvested in a shorter more predictable time period. Having a product and or procedure to induce leaf drop and start the dormancy period in storage would be of great interest to growers. Currently, the only product labeled for such use is Florel. Other materials such as chelated copper or the cotton defoliant herbicide Dimethipin have been used but with variable results.

ISSUES AND OPPORTUNITIES FOR PGRS IN COMMERCIAL PRODUCTION OF ORNAMENTAL PLANTS

Joyce Latimer

Virginia Tech

INTRODUCTION

There are lots of ways to approach a topic like this. I have chosen to approach it with the question, "Is the relationship between ornamental plant production and environmental stewardship necessarily an adversarial relationship?" This particularly revolves around the industry's relationship with water.

The economic impact the industry surrounding the production and use of ornamentals is estimated at ~$149 Billion and supports over 1.9 Million jobs in the United States alone (Hall et al., 2005). Production of ornamentals, or you could argue the production of any agricultural crop, is composed of two relationships with water that relate to environmental stewardship:

1) Impacts of water on the production and use of crops
2) Impacts of production and use of crops on water

In other words, we need the water to produce the crops but our production impacts the water and therefore, the environment, around us. Seems simple, right?

IMPACTS OF WATER

Although we don't have data specific to the production of ornamental crops, we know that freshwater withdrawals in the U.S. in 2005 totaled 410,000 Million gallons per day and that 37% of that withdrawal was used for irrigation (Figure 1; USGS, 2005). Furthermore, 42% of that irrigation water was withdrawn from ground water sources which are rapidly becoming a limited resource. Surface water sources are less preferred for the production of ornamental plants as these sources are typically of lower quality and may contain a variety of contaminants that limit plant production.

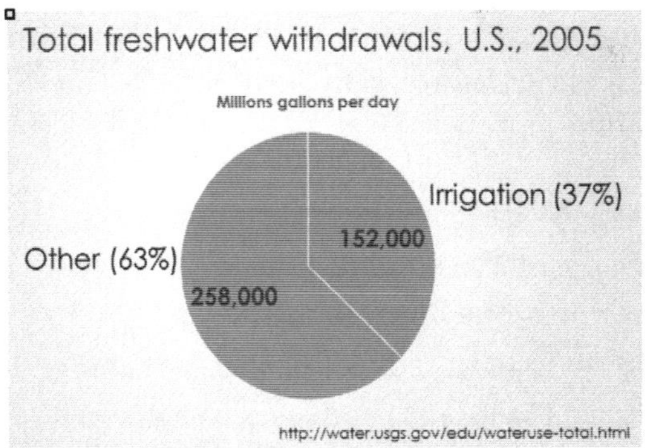

Fig. 1. Total freshwater withdrawal in U.S. in 2005. most recent data available (USGS. 2005).

Many areas of the U.S. have suffered moderate to severe drought conditions over the last 10 to 20 years and projections are that there will be more and greater areas of drought in the future (NRDC, 2010). In fact, the NRDC projects that, based on the levels of use and susceptibility to drought in 2010, we will have an even greater incidence of severe drought and reduced water supplies in the U.S. by 2050 (Figure 2) further limiting the availability of irrigation water.

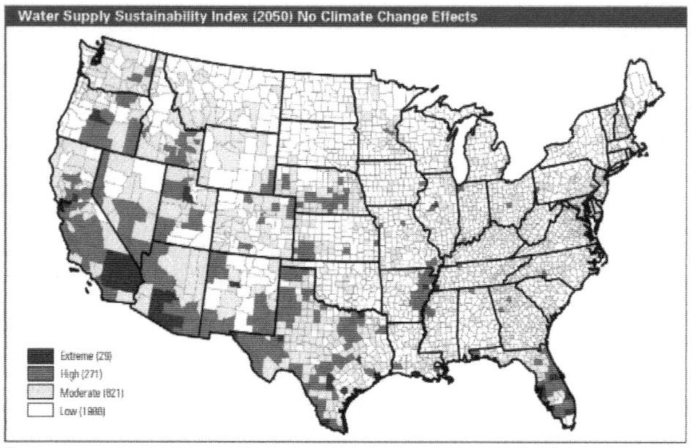

Fig. 2. NRDC projections of areas of low to extreme drought conditions in the U.S. in 2050 without factoring in climate change projections (NRDC, 2010). http://www.nrdc.org/globalWarming/watersustainability/

It has been estimated that there are about 241,500 acres of greenhouse and nursery production under irrigation in the U.S. (USDA, 2014). However, limited water availability for production is only part of the problem. As drought and subsequent irrigation restrictions move into urban and suburban areas, consumers and landscapers reduce or cease purchasing ornamental plants for their homes or commercial operations.

So drought puts pressure on consumers to reduce water use and to replace turf with drought resistant plants. Some city planners are offering rebates and cash incentives to homeowners who switch from turf to low maintenance, drought-tolerant ornamental plantings (McClellan, 2015). But perhaps, with the marketing issues we already have with younger consumers of ornamental plants, the biggest problem is that drought enhances their fear of failure. Why plant it to watch it die?

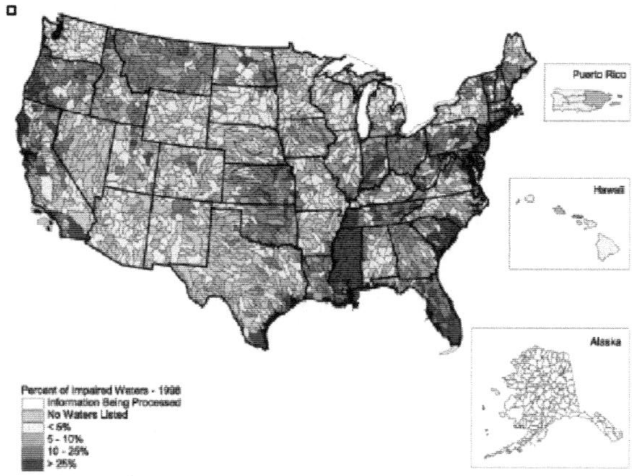

Fig. 3. Percent of impaired waters in the U.S. in 1998. Source: EPA http://nac.unl.edu/atlas/Map_Html/Clean_Water/National/Percent_impaired_water_nat/Percent_impaired_water_nat.htm

So we start the cycle of decline: since they have limited water, consumers buy fewer plants. This reduced market reduces production and we have the loss of producers. When the drought hit Georgia in 2007, Pike's Family Nurseries, one of the largest garden center chains in eastern U.S., was forced into bankruptcy. Georgia lost over 30,000 jobs in the green industry as a result of the drought (Hall, 2015). For the growers who survive droughts or prepare to face drought conditions in their production or market areas, there is significant pressure to conserve, collect and reuse their irrigation water. And, that brings us to the second issue – the impact of our industry ON water.

IMPACT ON WATER

Agriculture is considered a major culprit in the contamination and degradation of U.S. waterways. Runoff of chemicals, especially nutrients and pesticides, along with sediment has impaired many streams and subsequent waterways. A waterway is defined as impaired if one or more pollutants is present in greater than the Environmental Protection Agency's (EPA) prescribed amounts. As you can see in Figure 3, there are a lot of impaired waterways in the U.S. (EPA, no date). This graph depicts issues in 1998 but similar graphs posted by EPA for 2002 show little improvement (See http://www.fs.fed.us/ne/delaware/biotrends/trends77.JPG).

In our area, the restoration of the Chesapeake Bay is receiving the most attention. The Chesapeake Bay Watershed consists of parts of six states, a total of 64,000 square miles with about 11,600 miles of shoreline (Chesapeake Bay Program, 2015). Many of the watershed's 170 streams are on the EPA impaired waterway list. EPA mandated and approved cleanup plans for Virginia which include language directing all greenhouse and nursery operations to capture and reuse all of their operational water as well as up to 0.5 to 1.0 inches of storm water falling on their property (VaDEQ, 2012). While this is daunting to our small and mid-sized growers, many of our larger ornamental plant producers implemented recapture/reuse practices years ago, in part to protect themselves against water shortages in times of drought. In addition, some producers see water recycling as an environmental responsibility and have sought certification programs that guide and reward the process of becoming an environmentally sustainable production operation. And, in some cases these certifications have been required or "strongly suggested" by their customers, the buyers for major retail outlets.

So how does all this relate to PGRs? There have been some negative consequences of recycling greenhouse operational water. In larger operations where a significant amount of persistent growth retardants like paclobutrazol are used, these PGRs are accumulating in the recycled irrigation water. The Floriculture Research Alliance conducted a study of eight commercial greenhouses where they collected 95 samples of recirculated water from various collection basins to measure the concentration of paclobutrazol (Raudales et al., 2014). Seventy-eight (82%) of those samples contained greater than 5 ppb paclobutrazol while 65% of the samples had even higher rates, greater than 20 ppb. These rates affect growth of many of our bedding plants, especially when applied on a continuous basis. While researchers are looking at how to remediate these contaminated waters, there is no easy fix. A short term solution may be remediation of the plants using a gibberellic acid/benzyladenine product to overcome the growth retardation (Whipker, 2015). Several products are labeled for this use in the greenhouse.

And, there are other issues with PGRs. Growers have expressed dissatisfaction with our current PGR resources and the lack of predictability of plant response to PGRs in a variety of venues. These concerns and the many advancements in technology and automation have more companies developing trimming equipment for the greenhouse and more growers seeking the proper balance between PGR applications and automated trimming (Sparks, 2015). Overall, the use of PGRs is increasing, especially in the larger greenhouse operations as well as in nursery operations. Automated trimming may have niche applications but concerns with disease spread, such as the tobacco mosaic virus (TMV) outbreak on petunias last year encourages caution (Sparks, 2015).

OPPORTUNITIES TO MEET GROWER NEEDS

So, in summary, the issues are: Growers have to be concerned with water conservation as well as water protection. They also must increase their awareness of and in many cases, participation in, environmentally sustainability practices or programs. They already have some dissatisfaction with the PGRs available to them and have issues with PGR accumulation in the water that they are being mandated to recycle and reuse. So what are our opportunities to meet these grower needs?

NEW FORMULATIONS?

Although the grower still has to develop and use his knowledge of plant growth and development and experience with PGR applications and rates to effectively use PGRs, maybe we can eliminate some of the inconsistency issues by developing new formulations. In fact, there is already one success story, with paclobutrazol. Fine Americas, Inc. heard the complaints about how the activity of paclobutrazol products (a suspension formulation) was greater when solutions were mixed from product near the bottom of the container than from new containers, i.e., from the top of the container. The techs watched how the product was being handled and mixed. They realized that the larger (2.5 gallon or larger) containers were not being shaken sufficiently or regularly enough to keep the paclobutrazol in suspension. Therefore, over time, more of the active ingredient settled to the bottom of the container – and, yes, the activity of the solutions mixed at the end of a container was higher. So, Fine Americas developed a clear and more concentrated formulation of paclobutrazol and named it Piccolo 10 XC. The clear formulation has better solution stability. Yes, you still have to shake it, but since it is ten times more concentrated, Fine can sell it smaller containers that are easier to shake and the active ingredient doesn't settle like it did in the suspension formulation. According to the Fine Americas techs, once a grower tries Piccolo 10 XC, he typically stays with it. So, what other grower issues could we solve with formulation changes?

NEW USES?

We need to take the time to talk to growers, hear their issues and then to watch how they actually handle the products. That takes a lot of trust. Growers find their own ways of making things work. These methods may or may not be consistent with labeled uses. If they trust you enough to allow it, watch them. Watching how they use products, how they make applications, or how they streamline those applications for multiple crops, may help us rewrite labels that support those methods or it may feed back into changing the formulation or the development of new products altogether.

Fig. 4. Ethephon (Collate, Fine Americas, Inc.) applied to potted Nepeta 'Walkers Low' as a 50 ppm substrate drench (on right) at 10 fl. oz. per pot; untreated on left. Photo at 3 weeks after treatment (Latimer et al., unpublished).

Do we really know what works with the products we already have? How many years did we hear and report that ethephon has no soil activity? We took one man's work on foliar translocation and concluded that the product had no soil activity. Now we know different (Figure 4). Confirming ethephon drench activity effectively brings a new PGR product to market. However, did anyone ask the growers how they were applying ethephon in the greenhouse? If they had, we as researchers and industry reps would confirmed the drench activity of ethephon many years ago!

Then we have the recent report of daminozide activity in roots – liner soaks of petunia in relatively high concentrations of Dazide (daminozide, Fine Americas, Inc.) provided significant growth regulation (Figure 5; Whipker, 2015).

Fig. 5. 'Potunia Pink Pinkalicious' petunias were responsive to daminozide (Dazide, Fine Americas) when the rooted liners were soaked for 2 minutes in the PGR solution. Photo at flowering (Whipker, 2015).

So what do we have yet to learn about the other products we are already using? Are there combinations of our current products that could simplify decision-making or applications for the growers?

CAN WE PHYSICALLY IMPROVE PGR PREDICTABILITY?

What else can we do to improve PGR predictability? Can we use slow release or carrier technologies that are used in other industries, like pharmaceuticals? Can we develop a metered release comparable to multiple applications of small doses over time for continuous growth regulation? Would this application method provide more "forgiveness" in application techniques while providing effective growth regulation? What about polymer technology? Would polymer coatings improve product longevity or allow for more efficient applications? Can we improve product efficacy with coatings? Would a high tech coating reduce leaching of our PGRs, reduce our effect on runoff or recycled irrigation water? Or, maybe we can improve worker safety.

SUMMARY

We need to be open to new ideas and all the possibilities. The use of PGRs in ornamental plant production continues to increase but grower practices and values are changing. Although it might have been at one time, the relationship between our industry's production practices and environmental stewardship is no longer adversarial. Growers are increasingly expressing that water conservation and protection is their responsibility. Growers want and need product predictability. I propose that research and development specialists in both our universities and PGR companies have a responsibility to assist the growers in production and environmental stewardship. How can we, as an industry, help them grow better, more uniform crops more consistently and profitably while reducing their environmental impact?

LITERATURE CITED

Chesapeake Bay Program. 2015. Discover the Chesapeake Bay. Accessed August 27, 2015 at *http://chesapeakebay.net/discover*

Environmental Protection Agency (EPA). ND. Percent of impaired waters. Accessed August 27, 2015 at *http://nac.unl.edu/atlas/Map_Html/Clean_Water/National/Percent_impaired_water_nat/Percent_impaired_water_nat.htm*

Hall, C. 2015. The economic forecast. Greenhouse Management 2015 Structures Report, S12-S15.

Hall, C. R., A.W. Hodges and J.J. Haydu. 2005. Economic Impacts of the Green Industry in the United States. Final Report to the National Urban and Community Forestry Advisory Committee. Accessed August 27, 2015 at *http://www.ufei.org/files/pubs/EconomicImpactsoftheUSGreen%20Industr(NUCFACfinalreport).pdf*

McClellan, M. 2015. Feeling parched. Greenhouse Management June 2015 issue, pp. 52, 54, 56-57.

Natural Resources Defense Council (NRDC). 2010. Climate change, water, and risk: Current water demands are not sustainable. Water Facts Report. Accessed August 27, 2015 at *http://www.nrdc.org/globalWarming/watersustainability/*

Raudales, R., J. Altland, J. Boldt, L. Morris and P. Fisher. 2014. Paclobutrazol residues in recirculated irrigation water. Floriculture Research Alliance publication. Accessed July 15, 2015, *http://www.watereducationalliance.org/*

Sparks, B.D. 2015. Growers face dilemma in managing plant growth. Greenhouse Grower 33(6):60, 62.

U.S. Department of Agriculture (USDA). 2014. Farm and ranch irrigation survey (2013). 2012 Census of Agriculture Vol. 3, Special Studies part 1. Accessed August 27, 2015 at *http://www.agcensus.usda.gov/Publications/2012/Online_Resources/Farm_and_Ranch_Irrigation_Survey/fris13_3_041_041.pdf*

U.S. Geological Survey (USGS). 2005. Total water use in the United States, 2005. Accessed August 26, 2015; *http://water.usgs.gov/edu/wateruse-total.html*

Virginia Department of Environmental Quality (VaDEQ). 2012. Commonwealth of *Virginia Phase II Watershed Implementation Plan. Accessed August 27, 2015 at http://www.deq.virginia.gov/*

Whipker, B.E. 2015. Cutting edge PGRs. e-GRO Research Update #2015.06. Accessed August 27, 2015 at *http://e-gro.org/pdf/2015-06%20NCSU%20PGR%20Research.pdf*

PRE-TRANSPLANT LINER SOAK AND POST-TRANSPLANT SPRAY APPLICATIONS OF AUGEO INFLUENCE GROWTH AND DEVELOPMENT OF VEGETATIVELY PROPAGATED SINGLE-NODE PENNISETUM SETACEUM 'RUBRUM' CULM CUTTINGS

W. Garrett Owen[1] and Roberto G. Lopez[2]

[1] Department of Horticulture and Landscape Architecture, Purdue University, West Lafayette, IN 47907 USA

[2] Department of Horticulture, Michigan State University, East Lansing, MI 48823 USA

ABSTRACT

Purple fountain grass [*Pennisetum setaceum* (Forssk.) Chiov. 'Rubrum'] is a popular ornamental grass grown due to its aesthetically appealing dark purple foliage. Greenhouse growers propagate purple fountain grass from crown divisions, tissue culture plantlets, or culm cuttings. However, liners of purple fountain grass are often not well branched and tall when they are transplanted into the final container. Therefore, our objective was to evaluate and quantify the effectiveness of pre-transplant liner dips and post-transplant spray applications of benzyladenine (Configure®), dikegulac-sodium (Augeo®), or ethephon (Florel®) to increase the number of culms and reduce height of greenhouse-grown purple fountain grass. Single-node culm cuttings were harvested from stock plants and rooted in a glass-glazed greenhouse with an air temperature set point of 23 °C and DLIs of 4.0 and 9.0 $mol·m^{-2}·d^{-1}$, respectively. After 5 d in a finishing environment with an air temperature set point of 20.5 °C and DLI of 14.3 $mol·m^{-2}·d^{-1}$, liners were dipped for 2 min or sprayed with either 25, 50, or 100 $mg·L^{-1}$ benzyladenine, 200, 400, 800, 1,600, or 3,200 $mg·L^{-1}$ dikegulac-sodium, or 100, 200, or 400 $mg·L^{-1}$ ethephon, or deionized water, or pruned. Liners were transplanted into 12.7-cm round plastic containers. At 6 weeks after dip or spray applications, culm number (C_n), plant height, and plant dry mass was determined. In general, pre-transplant liner dip applications of 25 or 50 $mg·L^{-1}$ benzyladenine, 100, 200, or 400 $mg·L^{-1}$ ethephon, 200, 400, 800 or 1,600 $mg·L^{-1}$ dikegulac-sodium, water, or pruned plants resulted in similar C_n. However, liners sprayed with 50 $mg·L^{-1}$ benzyladenine or were hand pruned produced significantly more C_n than those that were sprayed with water. Increasing spray concentration of dikegulac-sodium resulted in similar C_n and provided greater height control compared to plants that were sprayed with water. Therefore, results indicate post-transplant liner spray applications may be effective methods to increase culm number and control plant height of single-node purple fountain grass culm cuttings.

INTRODUCTION

Ornamental grasses, especially those with aesthetically appealing dark foliage such as purple fountain grass 'Rubrum' have increased in popularity. Grown as a horticultural annual in the United States (U.S.) (Bailey, 1949; Drake, 1999), availability of young purple fountain grass plants or liners is problematic (Cunliffe et al., 2001) for growers due to seed sterility (Simpson and Bashaw, 1969) and loss of outdoor stock plants from killing frosts

(Cunliffe et al., 2001). Propagation of purple fountain grass is limited and divisions have been the recommended method (Wang et al., 1999) in recent years. In addition, growers have reported that crown divisions are more costly, while tissue cultured plantlets are more cost effective (M. Goyette, personal communication). Growers who receive divisions or tissue cultured plantlets during winter months and early spring have reported slow growth and delayed flowering, thus resulting in undesirable crops.

During production, purple fountain grass plants are pruned to hasten culm (shoot) development and branching and control height. Although this method is commonly practiced among the greenhouse industry, it is often time consuming, labor intensive, and increases the opportunity for disease entry and spread. Alternatively, plant growth regulators (PGRs) provide numerous benefits to growers such as controlling plant height, reduce stem elongation, or to increase branching. Previous research and literature has established an extensive list of PGR liner dip and spray recommendations for commonly produced annual bedding plants and herbaceous perennials. Plant growth regulator dip or spray recommendations for ornamental annual and perennial grasses is limited. Previous researchers have investigated PGR dips or spray applications on fountain grass [*Pennisetum*; (Wang et al., 2000)], pampas grass [*Cortaderia*; (Sellmer et al., 2001; Padhye and Groninger, 2009)], sedge [*Carex*; Padhye and Groninger, 2009)], giant miscanthus [*Miscanthus*; (Padhye and Groninger, 2009; Latimer et al., 2011)], muhly grass [*Muhlenbergia*; (Padhye and Groninger, 2009)], and feather reed grass [*Calamagrostis*; (Latimer et al., 2011)]. Additionally, many greenhouse growers seek cultural practices to increase the number of culms and control height of purple fountain grass. Therefore, the establishment of research and grower-based recommendations is needed for purple fountain grass.

Our objectives were to evaluate and quantify the effectiveness of 1) pre-transplant liner dip and 2) post-transplant liner spray applications of benzyladenine, dikegulac-sodium, or ethephon to 1) increase the number of culms and 2) reduce the height of vegetatively propagated single-node purple fountain grass 'Rubrum' culm cuttings.

MATERIALS AND METHODS

Stock plant culture

Stock plants of purple fountain grass 'Rubrum' were grown in a glass-glazed greenhouse at Purdue University in West Lafayette, IN (lat. 40° N) with exhaust fan and evaporative-pad cooling, radiant hot water, and retractable shade curtains controlled by an environmental control system (Maximizer Precision 10; Priva Computers Inc., Vineland Station, ON, Canada). The greenhouse day and night air temperature set points were 23 °C. The photoperiod was a constant 16 h (0600 to 2200 HR) consisting of natural day lengths with day-extension lighting from HPS lamps (e-system HID; PARSource, Petaluma, CA) that delivered a supplemental photosynthetic photon flux density of 100 $\mu mol \cdot m^{-2} \cdot s^{-1}$ at plant height [as measured with an amplified quantum sensors (LI-190SL; LI-COR Biosciences, Lincoln, NE)]

Cutting propagation and culture

On 13 February 2015 (Expt. 1) and 27 March 2015 (Expt. 2), 180 uniform 20-cm single-node culm cuttings were harvested from stock plants. The basal end of each cutting was briefly dipped into a solution containing 1000 $mg \cdot L^{-1}$ indole-3-butyric acid (IBA) + 500 $mg \cdot L^{-1}$ 1-napthaleneacetic acid (NAA; Dip'N Grow Liquid Rooting Concentrate; Dip'N Grow, Clackamas, OR) and individually placed into a 50-cell liner tray (70-mL individual cell vol.; T.O. Plastics, Clearwater, MN) filled with a commercial soilless medium comprised of (by vol.) 65% peat, 20% perlite, and 15% vermiculite (Fafard 2; SunGro Horticulture, Agawam, MA) and 50% course perlite (Strong-Lite Coarse Perlite; SunGro Horticulture). Cuttings were sprayed to runoff with a solution containing 300 $mg \cdot L^{-1}$ non-ionic surfactant (CapSil; Paulsboro, NJ) so water will not accumulate on the plant foliage.

All cuttings were rooted in a glass-glazed greenhouse under ≈56 % shade (Solaro 5620 O R FR; Ludvig Svensson, Inc., Charlotte, NC) under ambient daylight supplemented with a *PPF* of ≈57.6 ± 8.1 $\mu mol \cdot m^{-2} \cdot s^{-1}$ at cutting height [as measured with an amplified quantum sensor (LI-250A Light Meter; LI-COR Biosciences)], respectively, delivered from high-pressure sodium lamps from 0600 to 2200 HR (16-h photoperiod) to achieve a daily light integral (DLI) of 4.0 (Expt. 1) or 9.0 $mol \cdot m^{-2} \cdot d^{-1}$ (Expt. 2). The propagation bench surface was heated to maintain a constant rooting medium temperature of 26 °C. The greenhouse day and night air temperature set points were 23 °C and maintained a relative humidity (RH) set point of 70 %. Propagation environmental measurements were recorded every 30 s and averages were logged every 15 min by a data logger (Model CR1000; Campbell Scientific, Inc., Logan, UT). Propagation environmental data for Expt.1 and 2 are reported in Table 1. Every 10 min, 6 s of acidified water was applied as mist. Acidified tap water was supplemented with a water-soluble fertilizer (Jack's LX 16N–0.94P–12.3K Plug Formula for High Alkalinity Water, J.R. Peters, Inc., Allentown, PA) providing (in $mg \cdot L^{-1}$) 50 N with each misting event. After 28 d, rooted cuttings were removed from the propagation environment and were toned for 5 d.

Experiment 1

On 17 March, rooted cuttings were dipped into trays containing 0.5 L of solution containing 25, 50, or 100 $mg \cdot L^{-1}$ benzyladenine (Configure; Fine Americas, Inc.), 200, 400, 800, 1600, or 3200 $mg \cdot L^{-1}$ dikegulac-sodium (Augeo; OHP, Inc.), 100, 200, or 400 $mg \cdot L^{-1}$ ethephon (Florel; Lawn and Garden Productions, Inc.), or deionized water for 2 min or pruned 10-cm from the medium surface. The day following chemical or pruning treatment, liners were transplanted into 12.7-cm (885 mL) diameter plastic containers (ITML Horticultural Products) filled with a commercial soilless medium comprised of (by vol.) 65% peat, 20% perlite, and 15% vermiculite (Fafard 2; SunGro Horticulture). Plants were grown in a polyethylene-covered greenhouse with exhaust fan and evaporative-pad cooling and radiant hot-water heating controlled by an environmental computer (Wadsworth EnviroSTEP; Wadsworth Control Systems, Inc., Arvada, CO). Plants were grown at day and night temperature set points of 21 °C and under natural day lengths. Two amplified quantum sensors (SQ-212; Apogee Instruments, Inc., Logan, UT) measured *PPF* every 30 s and the average of each sensor was logged every 15 min by a data logger (Watchdog 2800 Weather Station; Spectrum Technologies, Inc.). The average (mean ± SD) air temperature and daily light integral (DLI) for 6 weeks was 22.3 ± 1.7 °C and 17.1 ± 8.6 mol·m^{-2}·d^{-1}.

Six weeks after chemical or pruning treatment, culm number [C_n; (determined by counting the number of developed culms)] and plant height (measured from the container rim to the tallest arching leaf blade) were determined and recorded. Plants were destructively harvested at the medium surface, individually bagged, dried in an oven at 70 °C for 1 wk, and plant dry mass (DM) was determined.

The experiment was a completely randomized block design in a factorial arrangement, with chemical and mechanical treatments (13 levels) and individual plant replicates (6 levels) as factors and the experiment was replicated in time. Data were analyzed using SAS 9.2 (SAS Institute, Cary, NC). Comparisons among treatments were performed using Duncan's multiple range test at $P \leq 0.05$.

Experiment 2

Expect where indicated, procedures for Expt. 2 were as described for Expt. 1. On 28 April, rooted cuttings were transplanted and were grown for 12 d before receiving either a post-transplant chemical spray application or pruning as described in Expt. 1. Greenhouse environment, plant culture, data collection, experimental design, and statistical analysis were similar to those outlined in Expt. 1. The average (mean ± SD) air temperature and DLI for 6 weeks was 22.5 ± 1.7 °C and 23.6 ± 8.0 mol·m^{-2}·d^{-1}.

RESULTS AND DISCUSSION

Experiment 1

Similar C_n, plant height, and plant dry mass were determined for plants dipped in concentrations of 100, 200, and 400 mg·L^{-1} ethephon, water, and for the pruned treatment (Table 2). Compared to plants that were pruned, benzyladenine concentrations of 25, 50, and 100 mg·L^{-1} did not effectively increase C_n, however 100 mg·L^{-1} benzyladenine effectively controlled plant height. In general, C_n was similar between plants dipped in dikegulac-sodium concentrations of 200 to 800 mg·L^{-1}, ethephon concentrations of 100 to 400 mg·L^{-1}, water dipped, and pruned plants. However, dikegulac-sodium concentrations >800 mg·L^{-1} and benzyladenine at 100 mg·L^{-1} suppressed culm growth and development. Visually, phytotoxic effects were observed and increased when liners were dipped in dikegulac-sodium concentrations from 800 to 3,200 mg·L^{-1}. Compared to plants that were pruned, plants dipped in dikegulac-sodium concentrations increasing from 400 to 1,600 mg·L^{-1} decreased plant height by 38.3%.

Experiment 2

In general, comparable C_n were observed when plants were either sprayed with water or a PGR application with the exception of 50 mg·L^{-1} of benzyladenine (Table 3). Pruning purple fountain grass or a post-transplant spray application of 50 mg·L^{-1} benzyladenine resulted in 22.5% and 36.9% more culms, respectively, compared to plants sprayed with water. The C_n of plants sprayed with 50 mg·L^{-1} of benzyladenine and the pruned treatment were similar, therefore post-transplant liner sprays of 50 mg·L^{-1} of benzyladenine may be a viable alternative to pruning. The C_n of plants sprayed with 3,200 mg·L^{-1} of dikegulac-sodium decreased by 15.2% and 41.2% when compared to plants treated with water or pruned, respectively. Although not statistically different, it is important to note that plants sprayed with water exceeded an average height of 70-cm. Similar plant heights were observed when plants were sprayed with water or benzyladenine concentrations of 25 and 100 mg·L^{-1}. Therefore, pruning, and all the ethephon and dikegulac-sodium spray concentrations tested effectively reduced the height of purple fountain grass 'Rubrum'. Dry mass of plants sprayed with benzyladenine concentrations from 25 to 100 mg·L^{-1} was reduced by 14.7%, respectively. Dikegulac-sodium concentrations from 200 to 800 mg·L^{-1} resulted in similar plant dry mass. Dikegulac-sodium concentration of 3,200 mg·L^{-1} resulted in the smallest plant with a dry mass of 15.9 g.

CONCLUSIONS

In general, a post-transplant spray application of 50 mg·L^{-1} benzyladenine or pruning were effective methods to increase C_n and dikegulac-sodium or ethephon sprays effectively suppressed plant height without negatively influencing C_n of purple fountain grass. In general, we found plants with fewer culms were taller while plants with more culms were shorter. Therefore, growers should consider their desired finish product prior to liner dip or spray application. For growers that want an alternative to hand pruning, recommendations would be to trial pre-transplant liner dips and post-transplant liner spray application rates of 50 mg·L^{-1} benzyladenine, 200 to 400 mg·L^{-1} dikegulac-sodium, or 100 to 400 mg·L^{-1} ethephon to control plant height, not negatively affect culm number and hasten purple fountain grass production.

ACKNOWLEDGEMENTS

We gratefully acknowledge Rob Eddy, Josh Craver, Madeline Olberg, Rosario del Mario, Alyssa and Andrea Hilligoss, Kyle Martin, and David Myers for greenhouse and laboratory assistance. We thank Fred C. Gloeckner Foundation, Inc., Pleasant View Gardens, Four Star Greenhouse, OHP, Inc., and IR-4 for funding; Pleasant View Gardens for plant material; ITML for containers; Sun Gro Horticulture for growing medium, J.R. Peters for fertilizer; OHP, Inc. and Fine Americas, Inc. for chemicals.

LITERATURE CITED

Bailey, L.H. 1949. Manual of cultivated Plants. Macmillian Co. NY.

Cunliffe, B.A., M.H. Meyer, and P.D. Ascher. 2001. Propagation of *Pennisetum setaceum* 'Rubrum' from cuttings. J. Envrion. Hort. 19:1–3.

Drake, R. 1999. The color Encyclopedia of Ornamental Grasses. Timber Press. Portland, OR.

Latimer, J.G., J. Freeborn, and V. Groover. 2011. Liner dips for growth regulations of vigorous herbaceous perennials. Acta horticulturae. 153-157.

Padhye, S.R., and J.K. Groninger. 2009. Influence of benzyladenine, trinexapac-ethyl, or uniconazole applications on height and tillering of six ornamental grasses. HortTechnology 19: 737-742.

Sellmer, J.C., C.R. Adkins, I. McCall, and B.E. Whipker. 2001. Pampas grass responses to ancymidol, paclobutrazol, and uniconazole substrate drenches. HortTechnology, 11:216-219.

Simpson, C.E. and E.C. Bashaw. 1969. Cytology and reproductive characteristics in *Pennisetum setaceum*. Amer. J. Bot. 56:31–36.

Wang, S., W. Carlson, H. Gao, A. Cameron, and R.D. Heins. 1999. Cultivating purple fountain grass. Greenhouse Grower May:58–62.

Wang, S., W. Carlson, H. Gao, A. Cameron, and R.D. Heins. 2000. Cultivating purple fountain grass, p. 104–105. In. R.D. Heins, A. Cameron, and W. Carlson (eds.). Firing up perennials, 2000 ed. Meister Media, Willoughby, OH.

Table 1. Average (mean ± SD) rooting medium temperature, propagation tent and greenhouse air temperature, relative humidity, and daily light integral (DLI) during propagation of Experiment 1 (Feb. 2015) and 2 (March 2015) purple fountain grass 'Rubrum' culm cuttings.

	Rooting medium temperature (°C)	Air temperature		Relative humidity (%)
		Propagation tent (°C)	Greenhouse (°C)	
	Set points			
	26.0 °C	23.0 °C	23.0 °C	70.0 %
Experiment 1	25.0 ± 1.1	23.0 ± 1.6	22.7 ± 2.2	74.1 ± 13.6
Experiment 2	26.6 ± 1.7	22.5 ± 2.1	22.5 ± 1.7	75.8 ± 17.5

Table 2. Culm number (C_n), plant height, and plant dry mass of purple fountain grass 'Rubrum' six weeks after liners were dipped in solutions containing benzyladenine, dikegulac-sodium, ethephon, or deionized water for 2 min or pruned 10-cm from the medium surface.

Treatment	$mg \cdot L^{-1}$	C_n (No.)	Plant height (cm)	Dry mass (g)
Benzyladenine	25	56.6 de[z]	24.8 ab	5.4 bc
Benzyladenine	50	64.1 cd	23.2 abc	6.3 ab
Benzyladenine	100	49.1 e	19.0 d	3.7 de
Dikegulac-sodium	200	66.3 abc	21.5 bcd	5.3 bc
Dikegulac-sodium	400	74.8 abc	20.6 cd	4.4 cd
Dikegulac-sodium	800	79.3 a	18.4 d	3.2 e
Dikegulac-sodium	1,600	58.2 de	14.9 e	1.9 f
Dikegulac-sodium	3,200	29.1 f	14.4 e	1.4 f
Ethephon	100	69.0 abcd	24.1 ab	6.4 a
Ethephon	200	73.1 abc	24.4 ab	7.3 a
Ethephon	400	67.4 abcd	23.1 abc	7.0 a
Water	---	67.0 abcd	25.9 a	7.3 a
Pruned	---	77.6 ab	24.0 ab	7.0 a
Significance		***[y]	***	***

[z] Within-column means followed by different lower-case letters are significantly different by Duncan's multiple range test at $P \leq 0.05$.

[y] NS or *** indicates non-significance or significance at 0.001.

Table 3. Culm number (C_n), plant height, and plant dry mass of purple fountain grass 'Rubrum' six weeks after liners were sprayed with solutions containing benzyladenine, dikegulac-sodium, ethephon, or deionized water for 2 min or pruned 10-cm from the medium surface.

Treatment	mg·L^{-1}	C_n (No.)	Plant height (cm)	Dry mass (g)
Benzyladenine	25	61.0 bc[z]	71.5 ab	33.6 a
Benzyladenine	50	72.3 a	65.7 bcd	32.9 abc
Benzyladenine	100	58.7 bc	68.9 abc	29.3 bcde
Dikegulac-sodium	200	57.8 bc	66.0 bcd	28.0 de
Dikegulac-sodium	400	54.9 c	64.4 cd	28.4 de
Dikegulac-sodium	800	57.1 bc	60.1 de	25.4 ef
Dikegulac-sodium	1,600	60.5 bc	55.8 ef	23.2 f
Dikegulac-sodium	3,200	45.8 d	50.9 f	15.9 g
Ethephon	100	57.7 bc	64.9 bc	28.8 cde
Ethephon	200	59.7 bc	62.3 d	33.5 ab
Ethephon	400	57.3 bc	62.8 cd	29.8 abcd
Water	---	52.8 cd	73.6 a	33.2 ab
Pruned	---	64.7 ab	68.9 abc	28.0 de
Significance		***[y]	***	***

[z] Within-column means followed by different lower-case letters are significantly different by Duncan's multiple range test at $P \leq 0.05$.

[y] NS or *** indicates non-significance or significance at 0.001.

IMPROVING BRANCHING OF NURSERY APPLE TREES

Winfred P. Cowgill, Jr, Rebecca Magron, Mike Beese

New Jersey Agricultural Experiment Station, Rutgers, The State University

Terence Robinson

Cornell University

Apple tree liners produced in nurseries worldwide typically do not form adequate branches (feathers) in their growth cycle to suit the demand of commercial orchardists utilizing high density production systems of 1100 trees per acre or more. Feathered nursery trees are a critical component of most high-density apple planting systems including the Tall Spindle. Six plant growth regulator research trials were conducted from 2012-2014 in consecutive years at a commercial nursery in the state of Delaware, USA, to identify plant growth regulator (PGR) materials and their application techniques to induce apple tree branching in the nursery. Various rates and timings of Maxcel® (6-Benzyladenine 1.8%) and Promalin® (6-Benzyladenine 1.8%) + Gibberellin A4 and Gibberellin A7) as compared to Tiberon™ (Cyclanilide (2.8%); 1,2-Propanediol were evaluated in 2012. Cultivars treated included Golden Delicious, cv 'Gibson' and Macoun. In 2013 an experiment was conducted on Fuji cv 'Daybreak' and Macoun to evaluate different rates and number of applications of Maxcel® or Promalin®. Our results in these 4 experiments indicate that both 4-5 applications of Maxcel® or Promalin® at 400-500 ppm significantly improve branching over Tiberon™ or the untreated control liners; with 10-17 branches on the Maxcel® and Promalin® liners vs 0-5 branches in the untreated control (UTC) or Tiberon™ liners. In 2014 two additional experiments were conducted to determine the effect of the number of Maxcel® or Promalin® sprays on branching of Fuji cv 'Daybreak' apple nursery liners and to screen wide range of additional varieties (12) for the best treatment, Maxcel® or Promalin®. In the 2014 Fuji experiment both Maxcel® and Promalin® induced a significant increase in the number of feathers with Fuji/M.9 trees compared to the untreated controls. Maxcel® treated trees had an average of 15 feathers while Promalin® treated trees had an average of 13 feathers and the untreated trees had 5 feathers. For the variety screening experiment we selected the best two treatments from our work in 2012 and 2013, Maxcel® @400 ppm and Promalin® @ 400 ppm + Surfactant (Regulaid® @1 pint/100gal). We also added treatment Promalin® @500 ppm + Surfactant. Among varieties, Cripps Pink (Pink Lady) and Enterprise had the highest tree quality ratings when treated with either Maxcel® or Promalin®. Aztec Fuji and Ambrosia had the next highest quality ratings followed by Crimson Crisp, Empire, Gala, Golden Delicious, Honeycrisp and Suncrisp. Royal Cortland, Ruby Mac had the least response to Maxcel® or Promalin®. There was no significant difference in tree quality (number of feathers and height) between those treated with Maxcel® or Promalin® when averaged over all 12 varieties. Standard recommendations for east coast USA nurseries are being developed from this work for the use of Maxcel® and/or Promalin® to allow for well-feathered apple liners in the nursery.

HOMOBRASSINOLIDE FOR THE PRODUCTION OF SHIPPING QUALITY OF SWEET CHERRIES

N. Bhushan Mandava, Repar Corporation, 8070 Georgia Avenue, Silver Spring, MD

Yan Wang and Todd Einhorn, Mid-Columbia Agricultural Research and Extension Center, Oregon State University, Hood River, OR

Scott Hicks, Innovative Ag Research and Consulting, LLC, 4741 W. Spruce Ave Suite 103, Fresno, CA

Dennis Dunbar, R3 Ag Consulting, 1629 Pollasky Avenue, Suite 111, Clovis, CA

Homobrassinolide (HBR) belongs to a group of about 70 brassinosteroids that are ubiquitous to higher and lower plants. It is registered with U.S. Environmental Protection Agency. We are evaluating its growth-promoting effects on fruit crops including cherries.

Cherry growers are reluctant to apply GA_3 on blush and early-mid-maturing cultivars due to its delaying effects on fruit color development and harvest timing. The objectives of this study are to determine the effect of a pre-harvest spray of homobrassinolide (HBR) on fruit quality at harvest and after storage/shipping and maturation of 'Rainier' and 'Bing', and the response of fruit growth, fruit size, yield, and return bloom to HBR.

We noticed the following significant findings:

(1) HBR at 1 ppm applied at straw color stage increased fruit firmness (FF) of 'Rainier' (numerically) and 'Bing' significantly ($p < 0.05$) without delaying color development and maturation. HBR did not significantly affect fruit size, pedicel pull force, soluble solid content (SSC), or titratable acidity (TA) at harvest or after storage/shipping.

(2) HBR and GA_3 tended to reduce fruit bruising and pedicel browning incidences but not at a statistically significant level after 3 weeks of cold storage.

These results as well as the return bloom will be presented.

CAN THE EFFICIENCY OF NITRATE USE IN MAIZE AND OTHER CULTIVATED PLANTS BE IMPROVED BY CHANGING THE HORMONAL BALANCE OF THE TREATED PLANT?

Karl-Friedrich Bangerth

University of Hohenheim

Fertilization with nitrate is one of the most limiting but also most expensive factor in the production of cultivated plants like maize and "Nitrate use efficiency" (NUE) is frequently researched in order to limit or optimise the use of NO3 in plant production. The often excessive and thus expensive fertilization with NO3 is the result of various factors, e.g. application of too much NO3 at one time (to avoid even more expensive rather frequent fertilizations) all of which cannot immediately be taken up by the plant and the excess is washed out of the soil by rain etc. Another reason can be the low NO3 uptake capacity of the plant at a time with the same result; the excess applied NO3 is lost. Inefficient NO3 use can also be due to the low absorption of NO3 by the soil resulting in a rapid wash out of nutrient etc. For these and other reasons, NUE is seldom better than 50% (Hirel et al., 2007). These problems of low NUE can possibly be overcome by treating the plants in such a way that a high biomass/seed production is achieved when the plant's hormone production/interaction is altered instead of attempting to change the NO3 uptake and/or NO3 metabolism of the plant. Such an approach would also save expensive NO3 fertilizer and/or reduce NO3 losses to the wider environment. The present experiments were initiated to investigate these possibilities.

WATERMELON, TOMATO, AND LETTUCE RESPONSES TO REDUCED NITROGEN RATES AND A COMMERCIAL EXTRACT OF ASCOPHYLLUM NODOSUM

Robin E. Ross[1], Gerald E. Brust[2], Troy Miller[3]

[1] *Market Development Scientist, Acadian Seaplants Limited, 30 Brown Avenue, Dartmouth, NS, Canada. Telephone: 704-907-7010. E-mail: rer@acadian.ca*

[2] *IPM Vegetable Specialist, University of Maryland, 2005 Largo Rd, Upper Marlboro, MD, USA.*

[3]*Coastal Crop Consulting, 73 Paseo Hermoso, Salinas, CA, USA*

Reducing nitrogen fertility inputs may lead to reductions in run-off and leaching, and be financially advantageous to growers. This research examines effects of nitrogen applied at reduced rates alone and in combination with *Ascophyllum nodosum* seaweed extract on yield of lettuce in the Salinas and Yuma valleys and tomatoes and watermelons near the Chesapeake Bay. Treatments in watermelons included nitrogen applied at rates of 100 and 150 pounds/acre, with and without *A. nodosum* seaweed extract. Reducing nitrogen resulted in significantly lower yields compared to the standard nitrogen program (281 vs 233 lb/A) however the lower fertility with the addition of seaweed extract treatment showed yield results similar to those from the standard rate N plots (281 vs 273 lb/A)). There were no differences in yield between the standard nitrogen treatments (150 lb/A) and the reduced treatments (105 lb/A) in the tomato trial however adding *A. nodosum* improved yield by 20%, increased petiole nitrate levels, and reduced blossom end rot. Four lettuce trials examining a grower`s normal fertility program, with reduced nitrogen programs, and *A. nodosum* seaweed extract applied to each, demonstrated that reducing fertility significantly decreases lettuce head weight by 12%(936.7g vs 821.8g), however, adding *A. nodosum extract* to the reduced fertility program resulted in yields similar to the grower`s normal fertility program (936.7g vs 922.5g). This research demonstrates that *A. nodosum* seaweed extract may be a valuable tool to help growers maintain yields in reduced fertility programs.

EFFECT OF HOMOBRASSINOLIDE FOR YIELD INCREASES IN WALNUTS

N. Bhushan Mandava

Repar Corporation, 8070 Georgia Avenue, Silver Spring, MD

Homobrassinolide (HBR) is one of the active growth regulating substances among a group of more than 70 brassinosteroids that are naturally occurring in plant kingdom. HBR was evaluated on several crops including fruit and nut crops for crop yield increases and also for improving the quality of the treated crops. Previously, we reported the HBR effect on almonds for increasing the nut size and the numbers of nuts retained at harvest. We now report the HBR effect on walnuts.

Walnut is a fruit whose productivity depends on the proportion of fruit curdles in limiting weather conditions or management of pollen. HBR was tested in Central Chile on two walnut varieties (Serr and Chandler) by spraying it at 1, 10 and 50 ppm when the trees were in full bloom, and also with CPPU (Forchlorfenuron) (10 ppm CPPU in combination with 10 ppm HBR). The effects in fruit set and on the weight of the nuts were measured at harvest time.

We found that the fruit set in cv. Serr was significantly better with all treatments of HBR, including with CPPU. The increase in fruit set was improved from 32.4% to 45.9% compared to control. The fruit set in cv. Chandler was significantly better with treatment of HBR at 50 ppm which showed 13.5% increase over control.

The increase in weight in shell in cv Serr was from 11.4 to 16.4% in all treatments, whereas there is no effect of CPPU which is similar to control.

Exposure of HBR to cv Chandler showed an increase in weight in shell at 1 ppm and 10 ppm (16.9% and 29.6% increase, respectively) whereas HBR at 50 ppm and HBR plus CPPU did not show any increase in weight in shell which is similar to control.

The details of these results will be presented.

FOLIAR APPLIED UNICONAZOLE EFFECTS ON YOUNG APPLE TREES

D.E. Deyton, L.F. Fowler and C.E. Sams

Department of Plant Sciences, The University of Tennessee, Knoxville, TN, 37996-4561 USA

Uniconazole (UCZ) is a triazole compound that is a potent inhibitor of gibberellin synthesis. Numerous trials have demonstrated that UCZ reduces vegetative growth when applied as a soil drench or foliar spray. Foliar sprays of UCZ were shown to reduce apple (*Malus xdomestica*) tree vegetative growth by ≈ 50% of controls at 10 weeks after treatment (Curry and Jones, 1990). UCZ (Sumagic®, Valent BioSciences Corp.) received a supplemental label in the USA for production of fruiting vegetable transplants (Schnelle and Barrett, 2009). UCZ obviously affects plant growth, but research results on effects on plant morphology and physiological functions have varied. UCZ has been shown to affect leaf net photosynthetic (Pn), stomatal conductance (*g*), and transpiration (*E*) rates in some trials and not in others. Differences in trials are often associated with differences in UCZ concentrations tested, method or timing of application

An objective of this project is to characterize the effects of UCZ concentration applied as a single foliar spray on shoot growth, Pn, *g* and *E* of young apple trees. The relationship of uniconazole concentration and effects on many plant morphological and physiological responses and vegetative growth can often be described by asymptotic analysis. The results of this experiment are shown by asymptotic relationships of UCZ concentration to plant responses.

MATERIALS AND METHODS

We planted one-year-old 'Golden Delicious'/M.M. 106 trees in 38 L pots in late winter, pruned them to 15 cm above the graft union and applied UCZ treatments on 15 June when ≈ 12 new leaves had formed. The trees were sprayed with 20 ml of 0 (control), 100, 200, 300, 400, or 800 mg/L UCZ. Treatments were emulsified with 0.1% (v/v) Tween 20. The trees and pots were leaned to the side at spraying and soil and pots were shielded with plastic to minimize UCZ contact with soil. The trees were arranged in a randomized complete block design with eight replications. Trees were grown outdoors on white polyethylene mulch and with drip irrigation. Pn, *g*, and *E* rates of newly fully expanded leaves were measured before treatment, and newly expanded leaves measured on 6 July (21 DAT (days after treatment)), 6 August (52 DAT), and 25 September (102 DAT) during the first growing season. Data were collected on cloudless days, between 1100 HR and 1400 HR, using a portable infrared gas analyzer (ADC LCA-2; Analytical Development Co., Hoddesdon, U.K.) to measure Pn and *g* rates and a steady state porometer (Licor L-1600) for *E* rates. The sampled leaf was then removed, leaf area determined with a Licor LI-3000 area meter, dried, weighed and Specific Leaf Weight (SLW) determined. An additional graph shows the effect of UBZ concentration on total shoot elongation at the end of the first and second growing seasons. Response curves to UCZ concentration were generated by SigmaPlot 12.0 (Systat Software Inc., San Jose, CA). Pn data from the 300 mg/L treatment were generally outliers and were not included in graphs. Any statistical information provided was determined by using analysis of variance and regression (PROC GLM) of Statistical Analysis Systems software (SAS Inst., Cary, NC).

RESULTS AND DISCUSSION

Vegetative growth: As expected, a single spray of UCZ in our study reduced of vegetative growth the first year. An asymptotic (exponential decay type) relationship (R^2=0.97) existed between UCZ spray rate and total shoot elongation (Fig. 1). Shoot elongation was also asymptotically reduced the second year (R^2=0.94). Thus, UCZ is a very potent growth retardant, but is also persistent in the plant.

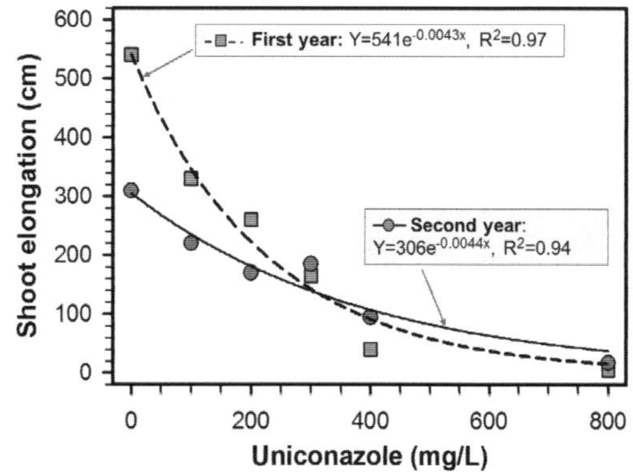

Figure 1. Total primary scaffold elongation of potted 'Golden Delicious'/ M.M.106 apple trees during the first and second growing seasons after treated with uniconazole sprays, n=8.

The second year reduction of shoot growth (compared to controls) on trees sprayed with 100 or 200 mg/L UCZ was less than the effect during first year, but higher spray rates still strongly retarded shoot elongation. The carryover effect on shoot growth necessitates that growers of perennial crops be aware and possibly adjust cultural practices.

Apple leaves developing after the UCZ treatment were influenced during the growing season. Specific leaf weight (SLW) of newly developing leaves was influenced by UCZ concentration ($P \leq 0.05$) throughout the growing season (Fig. 2). At 21 DAT, newly expanded leaves on UCZ treated trees had exponentially (R^2=0.99) more SLW as UCZ concentration increased. The relationship of UCZ concentration and SLW at 52 DAT can be described as linear (Y=10.1 + 0.009x, R^2=0.96) but a higher R^2 (0.99) existed when described by an exponential curve. A linear relationship of UCZ concentration and SLW existed at 102 DAT. However, the differences of SLW between leaves on untreated trees and leaves on trees treated with high UCZ concentrations were much greater at 102 DAT than earlier in the season. At 102 DAT, newly fully expanded leaves on trees treated with 400 or 800 mg/L were ≈ 35% to 45% smaller ($P \leq 0.05$) than leaves on unsprayed trees (data not shown). Previous research showed that three UCZ sprays (125 or 188 mg/L) increased apple SLW (Steffens and Zimmerman, 1992).

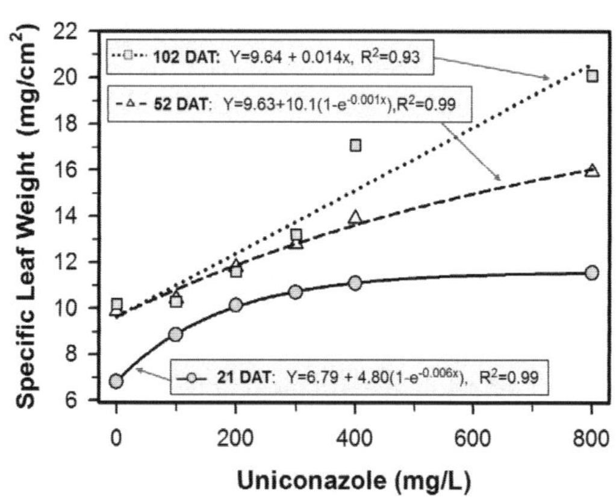

Figure 2. Specific Leaf Weights of newly fully expanded apple leaves at 21 DAT (days after treatment), 52 DAT and 102 DAT. Trees sprayed on 15 June with 0, 100, 200, 300, 400, or 800 mg/L uniconazole.

Photosynthesis: UCZ effect on leaf Pn rate was inconsistent in this study. An asymptotic relationship (R^2=0.86) existed between Pn rate (per unit of leaf area) of newly fully expanded leaves and UCZ concentrations at 21 DAT ($P \leq 0.05$, Fig. 3). Leaves on trees treated with 800 mg/L had ≈ 40% higher *A* rates than untreated trees. However, UCZ had no effect on Pn rate at 52 DAT. An asymptotic relationship (R^2=0.83) between UCZ concentration and Pn rate ($P \leq 0.05$) also existed late in the growing season (102 DAT, 25 Sept.). The increased photosynthesis late in the season may be beneficial, but also may increase tree susceptibility to freeze damage. These trees were over-wintered in a shade house and no cold damage was evident the next growing season.

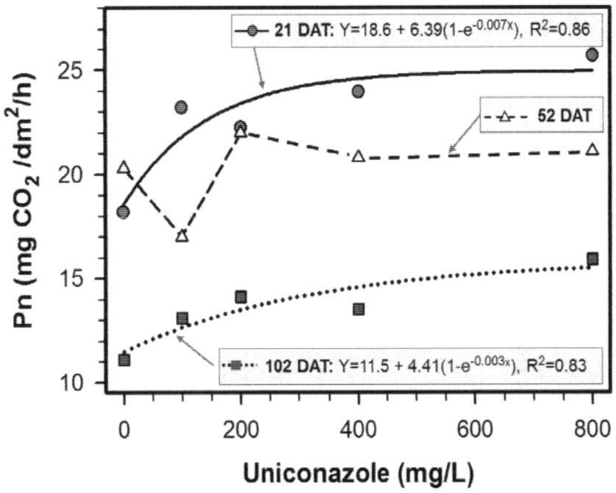

Figure 3. Net photosynthetic (Pn) rate of newly fully expanded apple leaves at 21, 52 and 102 DAT (days after treatment). Trees were sprayed on 15 June with 0, 100, 200, 300, 400, or 800 mg/L uniconazole.

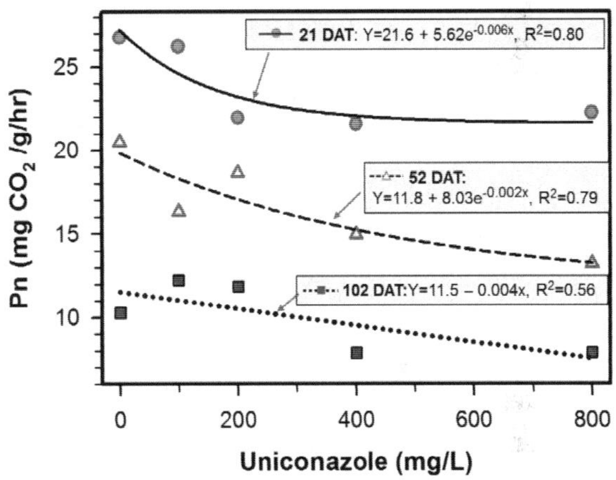

Figure 4. Net photosynthesis (Pn) per gram of dry leaf tissue of newly fully expanded apple leaves at 21, 52 and 102 DAT (days after treatment). Trees were sprayed on 15 June with 0, 100, 200, 300, 400, or 800 mg/L uniconazole.

Our research found that UCZ concentration had a negative effect ($P \leq 0.05$) on photosynthesis per unit of leaf dry weight (Fig. 4). Pn per gram of leaf tissue decreased exponentially at 21 DAT (R^2=0.80) and 52 DAT (R^2=0.79) with increasing UCZ rate. The significance of this finding is unclear. Photosynthesis per unit area or per gram dry weight of new leaves declined as the season progressed.

Reports of UCZ effect on Pn have been varied. Three sprays of UCZ (130 mg/L) did not significantly affect 'Haralred' apple leaf Pn (Saavedra et al., 1990b). Steffens and Zimmerman (1992) reported that three UCZ (125 or 188 mg/L) sprays increased mean seasonal Pn of apple leaves by about 10% and increased leaf chlorophyll concentration. Thetford et al. (1995) reported that forsythia (*Forsythia ×intermedia*) leaves developing after UCZ application had higher Pn rates and chlorophyll concentration on an area or fresh weight basis at 55, 77, and 365 DAT. They suggested the effect of UCZ on increased Pn rate may have been partially due to increased chlorophyll concentration. Sprays of low UCZ concentrations have been reported to also increase chlorophyll content in soybean (Zhang et al., 2007), wheat (Manal et al., 2010), barley (Hussein, 2014), and cotton leaves (Bekheta and Ramadan, 2005).

Stomatal Conductance and Transpiration: Stomatal conductance rates of newly fully expanded leaves in our experiment were generally higher on trees previously treated with higher UCZ rates (Fig. 5). Stomatal conductance rates were exponentially higher at 21 DAT (R^2=0.92, $P \leq 0.05$), 52 DAR (R^2=0.75), and 102 DAT (R^2=0.83) on trees sprayed with treated with higher UCZ rates. UCZ concentration affected transpiration rates of new leaves throughout the growing season in this study (Fig. 6). An asymptotic relationship existed between UCZ rate and transpiration rate on each measurement date; 21 DAT (R^2=0.96, $P \leq 0.01$), 52 DAT (R^2=0.76), and 102 DAT (R^2=0.93).

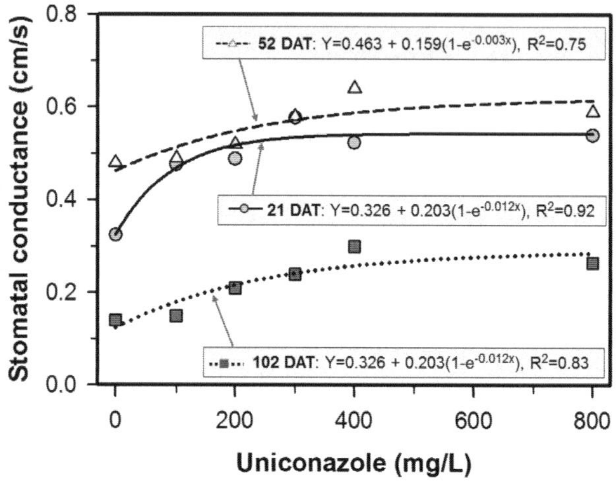

Figure 5. Stomatal conductance rates of newly fully expanded apple leaves at 21 DAT, 52 DAT, and 102 DAT. Trees were sprayed on 15 June with 0, 100, 200, 300, 400, or 800 mg/L uniconazole.

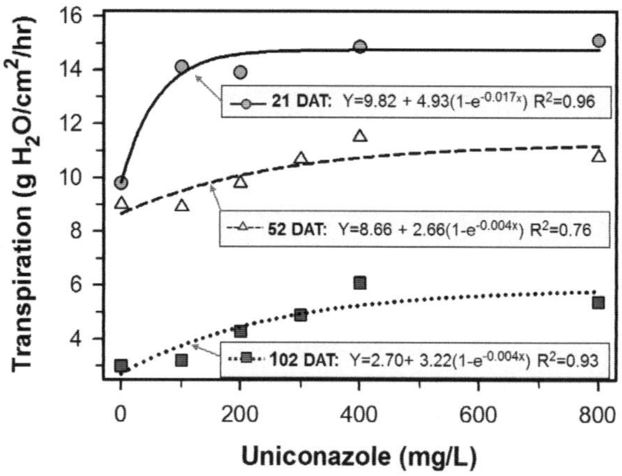

Figure 6. Transpiration rates of newly fully expanded apple leaves at 21 DAT, 52 DAT, and 102 DAT. Trees were sprayed on 15 June with 0, 100, 200, 300, 400, or 800 mg/L uniconazole.

Previous research of UCZ effect on stomatal conductance have reported differing results. Saavedra et al. (1990) reported that repeated foliar sprays UCZ (130 mg/L) did not influence *g*, *Pn*, or initial *E* rates of 'Roger McIntosh' apple leaves. Other research found that three UCZ (130 mg/L) sprays increased mean seasonal stomatal conductance of apple leaves (Steffens and Zimmerman, 1992). Thetford et al. (1995) reported that forsythia plants sprayed with UCZ (170 mg/L) had higher stomatal conductance rates at 49, 55, 77, and 365 DAT. They also found that stomatal density increased linearly with increasing uniconazole concentration, which could possibly affect leaf *g*, Pn or *E*. Steinberg (1991) reported that UCZ increased stomatal density of *Ligustrum japonicum* leaves, but that stomatal density or morphology did not necessarily influence *g* or *E* rates as soil water became limited.

Researchers have investigated UCZ treatment as a method to increase drought tolerance. We (Deyton et al. 2014) reported that apple leaf Pn rates of young trees sprayed with 100 to 400 mg/L UCZ were higher than control leaves for 15 days after imposed drought stress. Reduced water usage was reported for ligustrum plants (Steinberg et al., 1991) when UCZ was applied as a soil drench and for forsythia plants (Thetford et al., 1993) when applied as a foliar spray (170 mg/L). Seed treatments with UCZ increased drought tolerance of wheat (El-Khallal and Nafie, 2000) and corn (Batlang, 2006) seedlings. UCZ sprays at low concentrations (50 mg/L) have been reported to increase drought tolerance of soybean plants (Zhang, et al, 2007).

Various reasons have been proposed for the effects of UCZ on Pn, *g*, and *E*. It is well documented that UCZ reduces leaf area and increases root to shoot ratio, thus potentially reducing water stress. UCZ can increase stomatal density, but other factors appear to have greater influence on Pn, *g*, and *E*. Lakso (1983) demonstrated a close relationship of Pn and *g* rates for mature apple leaves under a wide range of water stresses. He reported that osmotic adjustment in mature leaves helped leaves maintain photosynthesis under water stress. Increases in abscisic acid (ABA) concentration frequently occurs when plants are exposed to stresses. Saito et al. (2006) reported that UCZ (50 mg/L) spray reduced ABA catabolism, increased ABA concentration by two fold and enhanced drought tolerance of *Arabidopsis*. Zhang et al. (2007) found that well-watered soybean plants sprayed at first bloom with UCZ (50 mg/L) had increased Pn and ribulose-bisphosphate carboxylase (RuBPCase) activity, and 26% less ABA concentration than watered control plants. They found that under water stressed conditions, UCZ treated plants maintained higher leaf water potential, Pn rates, and 68% more ABA than stressed control plants. UCZ treated plants also maintained more leaf chlorophyll content, RuBPCase, quantum yield of photosystem II, soluble sugars, proline; and promoted several antioxidant compounds.

Recent research showed that foliar sprays of UCZ (150 or 200 mg/L) on salinity stressed barley plants increased photosynthetic pigments, total plant carbohydrates and total soluble sugars compared to salinity stressed plants (Hussein et al., 2014). Bakheta and Hussein (2014) also reported that UCZ (150 or 200 mg/L) sprays to salinity stressed barley plants increased ABA, crude protein, total soluble protein and proline. They suggested the increase of proline, proteins and soluble protein may play a role in osmotic adjustment and thus in improving salinity tolerance.

Summary: Field trials of UCZ have often been exposed to many environmental influences that can result in large variability in data. Thus, it has frequently been difficult to show statistical differences unless large trials are conducted. The use of asymptotic analysis can often show treatment/effect relationships and aid in evaluating treatment effects.

The results in our trial show that UCZ treatments can increase PN, *g* and *E* rates through the growing season and affect shoot growth through the first and a second growing season. The higher UCZ rates are obviously too high for current production systems, but the response curves may aid in decision making for future trials. Because of the demonstrated effects on Pn, *g* and *E* rates in this trial and stress tolerance in other trials, evaluation of lower UCZ concentrations (e.g. <200 mg/L) on establishment and growth of young, non-fruiting fruit plants may be a viable cultural practice in the future. As water availability becomes more limited to future producers, application of plant growth regulators such as UCZ may become important for improving drought tolerance.

LITERATURE CITED

Bakheta, M.A., and Hussein, M.M. 2014. Uniconazole effect on endogenous hormones, proteins and proline contents of barley plants (Hordium vulgare) under salinity stress (NaCl). Bioscience 6:39-44.

Batlang, U. 2006. Studies with triazoles to alleviate drought stress in greenhouse-grown maize (*Zea mays*) seedlings. MS Thesis, Virginia Tech University, Blacksburg, VA.

Bekheta, M.A. and Ramadan, A.A. 2005. Influences of uniconazole and coumarin on some morphological, physiological aspects and fiber quality of cotton plants. J. Agric. Sci., Mansoura Univ., 30:3683-3696.

Curry, E.A. and Jones, K.M. 1990. Controlling growth of 'Delicious' apple trees with low-dosage applications of uniconazole or BAS111. J. Hort. Sci. 65:619-626.

Deyton, D.E., Fowler, L.F. and Sams, C.E. 2014. Morphological and physiological responses of young apple trees to foliar applied uniconazole. Acta Hortic. (ISHS) 1042:261-267

El-Khallal, S.M. and Nafie, E.M. 2000. Alleviation of drought damage for two cultivars of wheat seedlings by application of growth regulators brassinolide and uniconazole. Egyptian J. of Physiol. Sci. 24:297-317.

Lakso, A.N., 2003. Morphological and physiological adaptions for maintaining photosynthesis under water stress in apple trees. R. Marcelle et al. (eds.), Effect of Stress on Photosynthesis. Martinus NiJhoff/Dr. W. Junk Publishers, The Hague 1983.

Manal, F. M., Thalooth, A.T. and Khalifa, R.Kh.M. 2010. Effect of foliar spraying with uniconazole and micronutrients on yield and nutrients uptake of wheat plants grown under saline condition. J. of Amer. Sci. 6:398-404.

Saavedra, J. A., Stang, E.J. and Palta, J.P. 1990. Growth suppression in dwarf and semi-dwarf apple rootstocks by uniconazole. HortScience 25:1101.

Saavedra, J. A., Stang, E.J. and Palta, J.P. 1990b. First year effects of a GA synthesis inhibitor on growth, photosynthesis ethylene, freezing tolerance, and some fruit characteristics in apple trees. Proc. of the Plant Growth Regulat. Soc. of Amer., 17th Annual Mtg., St Paul, MN, USA. pp. 148-150.

Saito, S., Okamoto, M., Shinoda, S., Kushiro, T., Koshiba, T., Kamiya, Y., Hirai, N., Todoroki, Y., Sakata, K., Nambara, E., and Mizutani, M. 2006. A plant growth retardant, uniconazole, is a potent inhibitor of ABA catabolism in *Arabidopsis*. 2006. Bioscience Biotech. and Biochem. 70:1731-1739.

Schnelle, R. and Barrett, J. 2009. Sumagic and tomato transplants. Greenhouse Product News 19 (11).

Steffens, G.L. and Zimmerman, R.H. 1992. Influences of uniconazole on growth, fruiting, and photosynthetic activity of tissue culture-propagated own-rooted apple trees. Scientia Hort. 52:201-214.

Steinberg, S.L., Zajicek, J.M., and McFarland, Marshall J. 1991. Short-term effect of uniconazole on the water relations and growth of *Ligustrum*. J. Amer. Soc. Hort. Sci. 116(3):460-464.

Thetford, M., Warren, S.L, Blazich, F.A. and Ranney, T.G. 1993. Effects of uniconazole on photosynthesis, dark respiration, and water use efficiency of 'Spectabilis' forsythia. Proc. of Southern Nurserymen's Assn. 38:27-29.

Thetford, M., Warren, S.L, Blazich, F.A., and Thomas, J.F. 1995. Response of *Forsythia ×intermedia* 'Spectabilis' to Uniconazole. II. Leaf and Stem Anatomy, Chlorophyll, and Photosynthesis. J. Amer. Soc. Hort. Sci. 120:983-988.

Zhang, M., Duan, L., Tian, X., He, Z., Wang, B. and Li, Z. 2007. Uniconazole-induced tolerance of soybean to water deficit stress in relation to changes in photosynthesis, hormones and antioxidant system. J. of Plant Physiol. 164:709-917.

PROHEXADIONE-CA MODIFIES PHENOTYPE, ANTIOXIDANT LEVELS AND ENZYMATIC ACTIVITY ON JALAPEÑO PEPPER

Homero Ramírez[1], Lino J. Ramírez-Pérez[1], José H. Rancaño-Arrioja[2]

[1]*Departamento de Horticultura, Universidad Autónoma Agraria Antonio Narro, Calz. Antonio Narro 1923, Saltillo 25315, Coahuila, México.*

[2]*Dirección de Investigación, Universidad Autónoma Agraria Antonio Narro, Calz. Antonio Narro 1923, Saltillo 25315, Coahuila, México*

The use of growth retardants could be an alternative to improve yield and fruit quality of jalapeño pepper (*Capsicum annum* L.) growing in greenhouse; particularly when a high antioxidants fruit content is one of the main goals. Prohexadione calcium (P-Ca) is a bioregulator which reduces vegetative growth and increases antioxidants content in deciduous fruit species. The effect of P-Ca in jalapeño pepper is little known, therefore, the objective of this research was to evaluate the effect of P-Ca on the phenotype and fruit quality on jalapeño pepper hybrid Grande. The experiment was established at the Universidad Autonoma Agraria Antonio Narro in Saltillo, Coahuila, Mexico, between april and july 2012.

When seedlings had reached 10 true leaves, a first foliar application of P-Ca at 0, 100, 200 and 300 mg L^{-1} where conducted; and 15 days later, a second P-Ca application was conducted at the same dosages. A complete random statistical design was used, with four replicates per treatment. The results obtained were analyzed with the statistical program SAS for Windows version 9.1, and for the analysis of variance and comparison of means, the Tukey (P =.05) test was applied.

Prohexadione-Ca reduced the final height and diameter of main stem; as well as yield. The P-Ca provoqued ripen fruits with significative increases in the content of capsaicine, total carotenoids, vitamin C and catalase and peroxidase activity. Improvement in fruit quality compensates yield reduction since the benefit : cost (B:C) ratio in P-Ca fruits had a value of 0.46 as a result of a higher market price, whereas in control samples was 0.27. It is conclude that prohexadione-ca has a potential to improve fruit quality in jalapeño pepper hybrid Grande growing under greenhouse conditions.

PLANT GROWTH REGULATORS AFFECT BRANCHING AND FLOWERING IN THREE ECHINACEA CULTIVARS

Mara C. Grossman, Margaret Aiken, Holly Scoggins, Joyce Latimer

Department of Horticulture, Virginia Tech, Blacksburg, Virginia 24061

Keywords: benzyladenine, ethephon, dikegulac sodium, perennial

ABSTRACT:

Echinacea purpurea (L.) Moench, eastern purple coneflower, is a perennial plant in the Asteraceae family which is native to most of North America. This species, along with several others, has been utilized extensively in the hybridization of new, high-value cultivars. These hybrid cultivars vary significantly in their growth habits and requirements from the species. Foliar applications of plant growth regulators were applied in order to evaluate the effects on branching and flowering in hybrid *Echinacea* cultivars which have not yet been studied extensively. Liners of *Echinacea* 'Hot Papaya', 'Marmalade', and 'Harvest Moon' received two applications of PGRs, the first approximately one week after transplant and the second two weeks later. Treatments included: 400 mg·L^{-1} dikegulac sodium, 500 mg·L^{-1} ethephon, 600 mg·L^{-1} benzyladenine, and two experimental compounds. 'Marmalade' treated with benzyladenine had increased branches, shoot dry weight, and marketability rating at 13 weeks after initial treatment (WAIT) compared to controls. 'Hot Papaya' treated with benzyladenine had more branches than controls at 6WAIT; while there were no differences in branching when plants were finished at 14WAIT, benzyladenine increased the number of flower stalks (4 compared to 1.7 in controls). At 14WAIT, 'Harvest Moon' treated with benzyladenine had more basal branches than plants treated with dikegulac sodium but did not differ from controls. These three hybrid cultivars were responsive to benzyladenine, although their response to the other PGRs tested varied.

Introduction:

Echinacea purpurea (L.) Moench, eastern purple coneflower, is a perennial plant in the Asteraceae family which is native to most of North America, and is hardy in USDA hardiness zones 3-8 (Armitage, 2008). *Echinacea* is a top selling perennial in the U.S., popular with consumers for being a drought tolerant and low maintenance plant. A 2014 Greenhouse Management survey (Jancsurak, 2014), lists *Echinacea* as one of the top 15 plants grown by growers surveyed, with 30% of those polled planning to increase production of *Echinacea* in the coming year. Breeders began introducing new hybrid *Echinacea* cultivars to market beginning in 1997, developing many new flower forms and colors (Carey and Avent, 2014). These hybrid cultivars vary significantly in their growth habits and cultural requirements from the species; many branch poorly due to a tendency to flower prematurely (Pilon, 2011).

Spray applications of benzyladenine (BA) have been successful in increasing branching of *Echinacea*. Six *Echinacea* cultivars, when sprayed with 600 mg·L^{-1} BA (Configure, Fine

Americas, Inc., Walnut Creek, CA) two weeks after transplant, had significantly more branches than untreated controls at 4 weeks after treatment (Latimer and Freeborn, 2009).

The objective of this study is to evaluate the effects of foliar applications of plant growth regulators on branching and flowering in hybrid *Echinacea* cultivars which have not yet been studied extensively.

Materials and Methods:

Three hybrid *Echinacea* cultivars were studied: Hot Papaya, Marmalade, and Harvest Moon. Size 72 liners were obtained from commercial sources and transplanted into trade gallon containers (2.8 L) filled with a peat and pine bark media (Fafard 3B, SunGro Horticulture, Vancouver, CA). One week after transplant, the liners received one of the following spray applications: 400 $mg{\cdot}L^{-1}$ dikegulac sodium (DS, Augeo, OHP, Mainland, PA), 600 $mg{\cdot}L^{-1}$ benzyladenine (BA, Configure, Fine Americas, Inc., Walnut Creek, CA), 500 $mg{\cdot}L^{-1}$ ethephon (ETH, Collate, Fine Americas, Inc., Walnut Creek, CA), 250 $mg{\cdot}L^{-1}$ experimental compound 1 (EXP1), 100 $mg{\cdot}L^{-1}$ experimental compound 2 (EXP2), or untreated control (CONT). Treatments were repeated two weeks after the initial treatment (WAIT).

Plants were transplanted between May 6th and June 9th, 2014 and grown in a coldframe in Blacksburg, Virginia until they reached a marketable size, between 13 and 14 WAIT. Every two weeks, plant height, width, and number of branches were measured. At the end of the study, plants were rated for marketability, and flowering data and shoot dry weight measurements were taken. Data were analyzed using analysis of variance ($P \leq 0.05$, Student's t-test) JMP Pro 10 (SAS Institute Inc., Cary, NC).

Results:

The plant growth regulators tested affected branching and flowering in each of the three *Echinacea* cultivars. In 'Marmalade,' BA increased branches, shoot dry weight, and marketability rating in finished plants at 13 WAIT compared to controls (Table 1, Figure 1). Plants treated with DS, ETH, and EXP1 had fewer branches than controls. Flowering was uneven in this cultivar; only plants treated with ETH flowered more than 50%.

Table 1. Number of branches, percentage flowering, shoot weight, and marketability rating of Echinacea 'Marmalade' at 13 weeks after initial treatment (WAIT). Plants were treated with 400 mg·L^{-1} dikegulac sodium (DS), 600 mg·L^{-1} benzyladenine (BA), 500 mg·L^{-1} ethephon (ETH), 250 mg·L^{-1} experimental compound 1 (EXP1), 100 mg·L^{-1} experimental compound 2 (EXP2), or untreated control (CONT) one week after transplant. Treatments were repeated two WAIT.

PGR	No. of branches	% flowering	Shoot dry weight (g)	Marketability[y]
CONT	6.3b[z]	13	18.7b	2.0b
DS	2.9c	50	18.9b	2.1b
ETH	2.8c	63	22.4b	2.1b
BA	11.5a	25	27.3a	2.9a
EXP1	5.7bc	0	19.4b	2.0b
EXP2	9.0ab	13	18.2b	2.3b
P value	<0.0001	.0204	.0173	.0205

[z]Means within a column followed by the same letter are not significantly different (Student's t-test, P<0.05, n=8).

[y] Marketability scale: 1. Unmarketable. Severe stunting. Very poorly branched. 2. Not well branched. More media showing than covered. 3. Not flowering or flower stalks not upright. Well branched. 4. Flowering or budding. Well branched. Little to no media showing. Upright flower stalks. 5. Well branched. No media showing. Good color. Good flowering with multiple flowers.

*Figure 1. Number of branches of Echinacea 'Marmalade' treated with 400 mg·L⁻¹ dikegulac sodium (DS), 600 mg·L⁻¹ benzyladenine (BA), 500 mg·L⁻¹ ethephon (ETH), 250 mg·L⁻¹ experimental compound 1 (EXP1), 100 mg·L⁻¹ experimental compound 2 (EXP2), or untreated control (CONT) counted at 2 week intervals after the initial PGR application (WAIT). NS, * nonsignificant or significant at P≤0.05.*

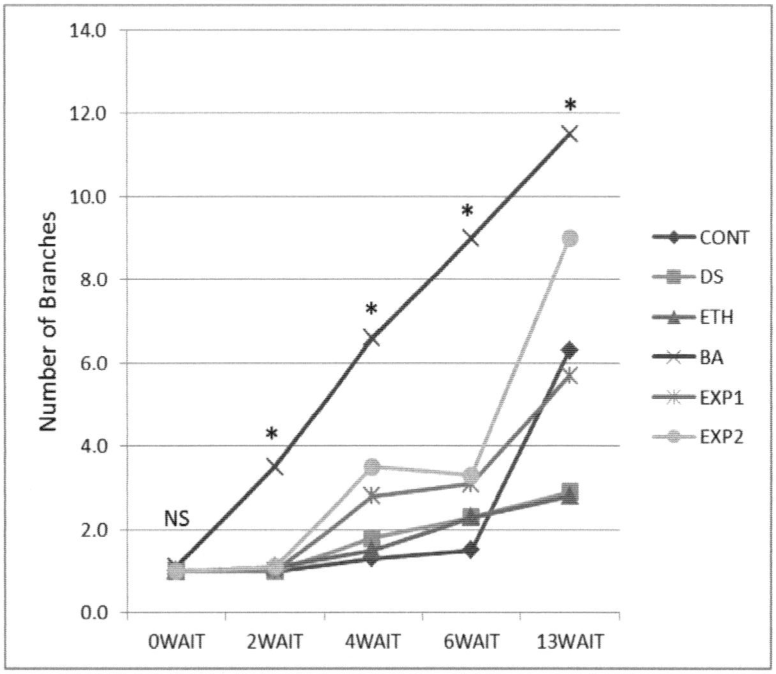

In 'Hot Papaya,' BA increased the number of branches in plants at 6 WAIT although there were no significant differences in branching at 14 WAIT (Table 2). In finished plants at 14 WAIT, BA increased the number of flower stalks, 4.0 compared to 1.7 in controls. EXP1 reduced the percentage of plant which flowered. Shoot dry weight was not affected by PGR application in this cultivar (data not presented).

Table 2. Number of branches at 6 and 14 weeks after initial treatment (WAIT) and percentage flowering and number of flower stalks of Echinacea 'Hot Papaya' at 14 WAIT with 400 mg·L⁻¹ dikegulac sodium (DS), 600 mg·L⁻¹ benzyladenine (BA), 500 mg·L⁻¹ ethephon (ETH), 250 mg·L⁻¹ experimental compound 1 (EXP1), 100 mg·L⁻¹ experimental compound 2 (EXP2), or untreated control (CONT).

PGR	No. of branches at 6 WAIT	No. of branches at 14 WAIT	% flowering	No. of flower stalks
CONT	4.8b[z]	6.5	83.3	1.7bc
DS	5.9ab	7.4	71.4	2.6bc
ETH	5.0b	6.6	87.5	2.1bc
BA	7.8a	9.5	87.5	4.0a
EXP1	6.3ab	7.5	50	1.0c
EXP2	5.9b	8.6	100	2.1bc
P value	0.0472	0.2086	0.1505	0.0020

[z]Means within a column followed by the same letter are not significantly different (Student's t-test, $P<0.05$, $n=8$).

In 'Harvest Moon,' plants treated with BA had more basal branches than plants treated with dikegulac sodium but did not differ from controls (Table 3). Percentage of plants flowering, shoot dry weight, and marketability were not significantly affected by treatments with PGRs. Two treatments were eliminated from the study due to infection with Impatiens Necrotic Spot Virus.

Table 3. Number of branches, percentage flowering, shoot dry weight, and marketability rating of Echinacea 'Harvest Moon' at 14 weeks after initial treatment (WAIT) treated with 400 mg·L⁻¹ dikegulac sodium (DS), 600 mg·L⁻¹ benzyladenine (BA), 250 mg·L⁻¹ experimental compound 1 (EXP1), or untreated control (CONT).

PGR	No. of branches	% flowering	Shoot dry weight (g)	Marketability rating
CONT	12.7ab[z]	83	23.9	3.5
DS	9.2b	67	20.7	3.5
BA	15.1a	86	26.1	4.0
EXP1	9.7b	71	24.3	3.0
P value	0.0041	0.8186	0.5668	0.1018

[z]Means within a column followed by the same letter are not significantly different (Student's t-test, $P<0.05$, $n=8$).

Discussion:

The three hybrid *Echinacea* cultivars were responsive to PGRs, although each responded somewhat differently. 'Marmalade' was the most responsive to BA, with significant increases in branches at every data measurement after treatment. 'Harvest Moon' and 'Hot Papaya' had increased branches with BA, although there were no significant differences in finished plants. 'Hot Papaya' or 'Harvest Moon' treated with DS, ETH, EXP1, or EXP2 did not significantly differ in number of branches as compared to controls. 'Marmalade' treated with DS or ETH had fewer branches than controls.

In 'Hot Papaya,' BA increased the number of flower stalks, while EXP1 decreased flowering. PGRs did not affect flowering in 'Harvest Moon.' Flowering was inconsistent in 'Marmalade' in this study.

These cultivars responded to foliar applications PGRs, with increased branching in response to BA over the course of production. Further studies, including studies on additional hybrid cultivars and studies evaluating growth retardants, would help us to understand the differences between the hybrid cultivars and their needs during production.

Literature Cited:

Armitage, A.M. 2008. Herbaceous perennial plants: A treatise on their identification, culture, and garden attributes. 3rd ed. Stipes, Champaign, IL.

Carey, D. and T. Avent. 2014. Echinacea explosion - the coneflower chronicles. Retrieved 14 September, 2014, from http://www.plantdelights.com/Article/Echinacea-Coneflower.

Jancsurak, J. 2014. Top 15 plants. Greenhouse Management. 34(1).

Latimer, J. and J. Freeborn. 2009. New uses of PGRs in ornamentals: Configure (6-BA) increases branching of herbaceous perennials. Plant Growth Regulat. Soc. Amer. 36:88-93.

Pilon, P. 2011. Tweak Your Growing Style to Succeed with Problem Perennials: Echinacea. Presented at OFA Short Course, July 9, 2011.

MECHANICAL STIMULI INDUCE EXPRESSION OF GENES INVOLVED IN DISEASE RESISTANCE IN ACACIA KOA

Kazue Ishihara[1], Eric Lee[1], Dulal Borthakur[1]

[1]Department of Molecular Biosciences and Bioengineering, University of Hawaii at Manoa, Honolulu, HI 96822

Acacia koa (koa) is a leguminous timber wood tree endemic to the Hawaiian Islands. Koa forests have been suffering from a devastating wilt disease caused by the fungal pathogen *Fusarium oxysporum* f. sp. *koae*. To protect the existing and future koa forests from the disease, it is crucial to identify genetic and environmental factors affecting disease susceptibility and resistance. The specific objectives of this study were to identify genes related to disease resistance and also to identify environmental factors that influence expression of those genes. In this study, Illumina-based transcriptome sequencing and Trinity *de novo* assembly yielded 85,533 unigenes with an average length of 641 bp. Using the Kyoto Encyclopedia of Genes and Genomes database, we identified complete coding sequences of 8 key genes involved in the plant-pathogen interaction pathways and 13 genes involved in the phenylpropanoid pathway, which is known to be associated with disease resistance. To further investigate links between those genes and disease resistance, we inoculated koa seedlings by *F. oxysporum* and are currently confirming if the genes are induced by the fugal infection through nCounter and qPCR analyses. Also, we are currently testing if the key genes are upregulated in response to mechanical stress since our results show that stressed seedlings are more resistant to *F. oxysporum* than unstressed seedlings with the survival rates of 47% and 18%, respectively. Exposure to wind during early development may induce resistance. The genes isolated in this study may also be used as potential molecular markers for selecting disease resistant koa, and it will contribute to genetic improvement programs for koa. This research is supported by the McIntire-Stennis Cooperative Forestry Program and a Monsanto Graduate Fellowship.

THE EFFECTS OF FOLIAR SYNTHETIC ABSCISIC ACID (S-ABA) ON CITRUS NURSERY TREES AND MATURE ORCHARD TREES

Lauren C. Garner and Adam Ingrao[1]

Horticulture and Crop Science Department, California Polytechnic State University, San Luis Obispo, CA 93407

ABSTRACT

Exogenous s-ABA applications can stimulate stomatal closure, enabling plants to better withstand short periods of drought by reducing transpirational water loss. To determine the effects of s-ABA on nursery trees, foliar applications of 0, 50, 100, or 150 mg·L^{-1} s-ABA were applied to potted 'Washington Navel' orange trees. A randomized complete block design with four blocks was utilized and the experiment was conducted twice. To determine the effects of s-ABA on mature orchard trees, foliar applications of 0 or 150 mg·L^{-1} s-ABA were applied to 20 year-old 'Olinda' Valencia orange trees. A randomized complete block design with five blocks was utilized. Potted citrus trees treated with s-ABA wilted 10-15 days after untreated control trees, had significant reductions in evapotranspiration as compared to untreated control trees and remained marketable significantly longer than untreated controls. No significant differences in leaf or fruit abscission were detected between control and treated orchard trees. Phytotoxicity was not observed during any of the experiments. Taken together, these results demonstrate that s-ABA could be utilized to extend marketability for nursery growers and might benefit citrus producers by reducing plant water use when water availability is limited.

INTRODUCTION

Post-production plant health is a major concern to both nursery producers and growers planting new trees and establishing orchards. For nursery growers, inventory loss due to drought stress during shipping and retailing of container plants is an important problem. At the point of shipping, nursery producers have made the largest capital investment in their crops and it is important to both nursery producers and growers that trees arrive without suffering drought stress and remain unstressed through post-planting irrigation. Exogenous applications of synthetic abscisic acid (s-ABA) have been shown to extend the marketable life of several ornamental crops (Barrett and Bartuska, 2011; Waterland et al., 2010) by signaling stomatal closure, thereby slowing plant transpiration rates. It may also be possible to extend the marketable life of woody nursery trees, such as citrus, with exogenous s-ABA applications. However, anecdotal evidence suggests that citrus trees are susceptible to phytotoxicity in response to exogenous ABA applications. The objectives of this project were to determine if s-ABA applications (1) reduce evapotranspiration and extend marketability of potted citrus trees in a post-nursery production environment and (2) result in phytotoxic effects in mature citrus orchards.

MATERIALS AND METHODS

Potted Citrus Tree Experiment

Using a randomized complete block design with 4 blocks, each experimental unit (a potted 1-year-old 'Washington Navel' orange on C-35 citrange) was treated with 0, 50, 100, or 150 mg·L^{-1} s-ABA as a foliar spray applied until runoff. Trees were maintained in a climate-controlled greenhouse at Cal Poly in San Luis Obispo, CA and were surrounded by well-watered non-study trees to mitigate any edge effect (Fig.1). The experiment was conducted twice, with Trial 1 from 15 Dec 11 to 29 Jan 12 (35/16 °C mean daily maximum/minimum greenhouse temperature) and Trial 2 from 3 April 12 to 4 June 12 (37/16 °C mean daily maximum/minimum greenhouse temperature). Plants remained unwatered for the duration of the experiment and observations were collected between 0800 and 1000 daily until all blocks were unmarketable. Evapotranspiration was measured by weighing each experimental unit daily. Wilting and chlorosis were measured using categorical scales.

Fig. 1. 'Washington Navel' nursery trees treated with 0, 50, 100, or 150 mg·L^{-1} s-ABA and maintained unwatered in a greenhouse.

Mature Orchard Tree Experiment

Using a randomized complete block design with 5 blocks, each experimental unit (a 20 year-old 'Olinda' Valencia orange on C-35 citrange) was treated with 0 or 150 mg·L^{-1} s-ABA as a foliar spray applied until runoff. The experiment was conducted once between 14 Dec 13 and 8 Jan 14 at the Cal Poly orchards in San Luis Obispo, CA. Trees were observed daily for fruit abscission and visual scales were used to evaluate leaf abscission and chlorosis in each of 4 quadrants of each tree.

Statistical Analyses

Data was analyzed by multivariate analysis of variance performed with JMP 11.2 (SAS Institute, Inc. 2014). Treatment means were separated using Tukey's honestly significant difference test with experimentwise error rate at $\alpha = 0.05$. Temporal profiles of relative weight loss were analyzed using Wilks' Lambda Test.

RESULTS AND DISCUSSION

Percent weight loss due to evapotranspiration was significantly affected by s-ABA treatments (Fig. 2; $P < 0.001$). Throughout Trial 1, percent weight loss was significantly higher in control potted citrus trees as compared to each s-ABA treatment. Throughout the marketable period of Trial 2, significant differences in percent weight loss were detected between the control and 100 mg·L^{-1} s-ABA treatments. Differences in percent weight loss did not vary significantly between the three s-ABA treatments during either trial.

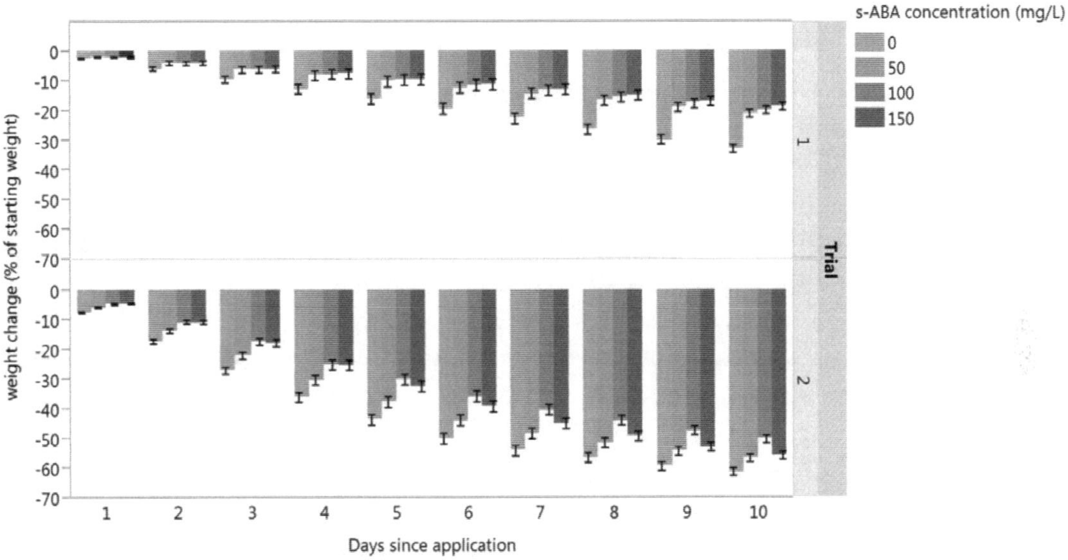

Fig. 2. Effects of s-ABA on daily weight change as a percent of initial weight of unwatered potted 'Washington Navel' orange trees during two trials. Error bars are model-based SEs used for comparing group means.

Control potted citrus trees wilted more quickly than those treated with s-ABA (Fig. 3) and wilting scores were significantly affected by s-ABA treatments ($P < 0.001$). During Trial 1, wilting scores were significantly different between control trees and those treated with s-ABA. There were no significant differences in wilting scores between the three s-ABA treatments.

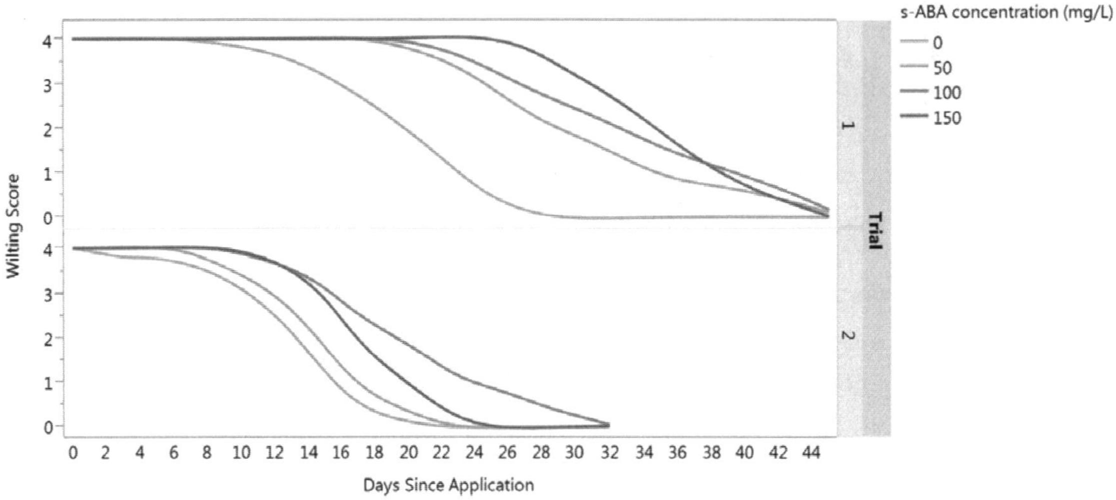

Fig. 3. Effects of s-ABA on wilting of unwatered potted 'Washington Navel' orange trees during two trials. Leaf wilting scale: 4 = full turgor; 3 = soft to the touch, 2 = beginning to wilt, 1 = completely wilted, 0 = dry and brittle.

Days to unmarketability was significantly affected by s-ABA treatments (Fig. 4; $P < 0.001$). During Trial 1, s-ABA treatments significantly increased days to unmarketability of potted citrus trees but significant differences were not detected between the three s-ABA treatments. Treatment differences were not detected during Trial 2.

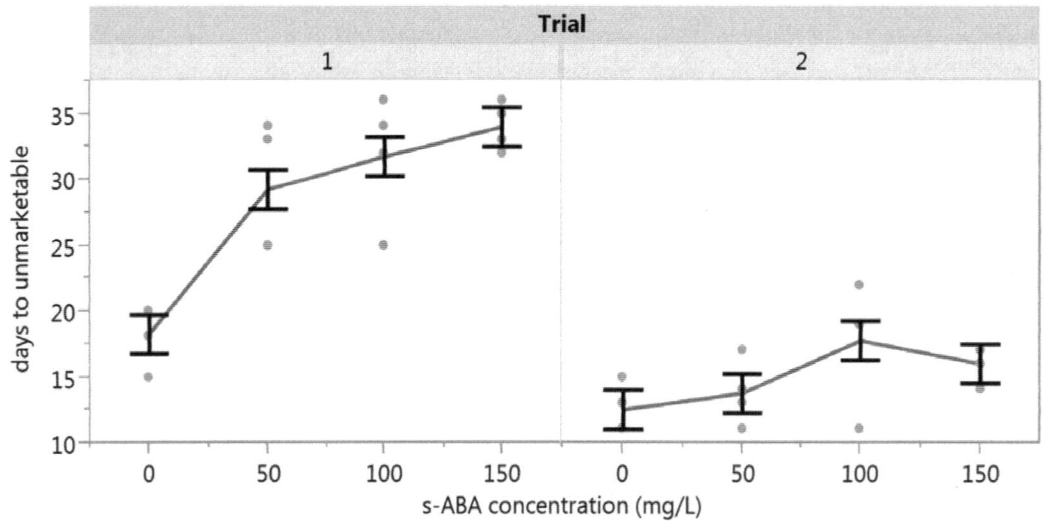

Fig. 4. Effects of s-ABA on the number of days before unwatered potted 'Washington Navel' orange trees became unmarketable during two trials. Error bars are model-based SEs used for comparing group means.

No significant differences in fruit or leaf abscission were detected between control orchard trees and the 150 mg·L^{-1} s-ABA treatment (data not shown).

Phytotoxicity was not detected during the greenhouse or field experiments.

In conclusion, foliarly applied s-ABA decreased evapotranspiration, increased days to wilting, and increased number of days that unwatered potted citrus trees remained marketable. Overall, s-ABA offers a viable option to extend marketability and plant health of citrus nursery trees during post-production, and at the doses reported herein, offers citrus growers a potential tool for mitigating drought stress without risk of phytotoxicity.

LITERATURE CITED

Barrett, J.E. and C.A. Bartuska. 2011. S-ABA delays wilting in petunia and impatiens. Acta Hort. 893:1271–1276.

Waterland, N.L., C.A Campbell, J.J. Finer, and M.L. Jones. 2010. Abscisic acid application enhances drought stress tolerance in bedding plants. HortScience. 45:409–413.

HYDROGEN SULFIDE MODULATES CADMIUM-INDUCED PHYSIOLOGICAL AND BIOCHEMICAL RESPONSES TO ALLEVIATE CADMIUM TOXICITY IN RICE

Mohammad G. Mostofa[1, 2], *Anisur Rahman*[1], *Ayaka Watanabe*[1], *Masayuki Fujita*[1]

[1]*Laboratory of Plant Stress Responses, Department of Applied Biological Science, Faculty of Agriculture, Kagawa University, Miki, Kagawa 761-0795, Japan*

[2]*Department of Biochemistry and Molecular Biology, Bangabandhu Sheikh Mujibur Rahman Agricultural University, Gazipur 1706, Bangladesh*

INTRODUCTION

Cadmium (Cd) is a potential and persistent environmental contaminant, causing serious toxicity to all living organisms including humans and plants. Cd has been ranked 7th among the top 20 toxins and considered human carcinogen (Gill and Tuteja 2011). In many South and East Asian countries, including Bangladesh, India, Japan, Indonesia and Thailand, Cd accumulation in rice and its subsequent transfer to the human food chain is a major environmental issue (Bolan et al. 2013). Rice contributed to the 36–50% of the total oral intake of Cd for Japanese population during 1998–2001 (Kikuchi et al. 2008). In Bangladesh, rice cultivating lands adjacent to the industrial establishments are highly contaminated with Cd that was found to be between 134 and 156 mg Cd Kg^{-1} of soil (Naser et al. 2009). Thus, preventing Cd uptake in rice plants grown in Cd-contaminated soils has become an urgent task to ensure food safety.

Cd, a non-redox water soluble heavy metal, can be quickly taken up by plant roots, and transported to the aerial parts where it significantly impedes vital cellular processes, including respiration and CO_2 fixation (Gill and Tuteja 2011). At the cellular level, Cd can bind to sulfhydryl and carbonyl groups of proteins and can replace essential cofactors, resulting in enzyme inactivation and production of reactive oxygen species (ROS), leading to oxidative stress induced damage (Ortega-Villasante et al. 2005; Gill and Tuteja 2011). Additionally, Cd is known to induce the production of methylglyoxal (MG), an extremely reactive aldehyde that can modify proteins, nucleic acids and carbohydrates by forming cross linkages (Hossain et al. 2009; Kaur et al. 2014). Moreover, MG in excess can further intensify ROS production by inactivating antioxidant enzymes or by interfering with photosynthesis (Saito et al. 2011; Hossain et al. 2012).

Plants have developed several strategies to counteract Cd toxicity, among which sulfur-induced defense is of paramount importance (Gill and Tuteja 2011; Khan et al. 2015). Sulfur-rich small ligands, such as cysteine, glutathione (GSH) and phytochelatins, directly bind with Cd and sequestrate it into vacuoles (Gallego et al. 2012; Seth et al. 2012). In addition, plant cells employ a number of GSH-based antioxidative reactions to scavenge Cd-induced superoxide ($O_2^{\cdot-}$), hydrogen peroxide (H_2O_2) and hydroxyl radical ($OH^{\bullet}$) (Hossain et al. 2012). GSH, with the help of ascorbate (AsA), plays central role in operating the AsA-GSH cycle in which ascorbate peroxidase (APX) detoxifies H_2O_2 in coordination to monodehydroascorbate reductase (MDHAR), dehydroreductase (DHAR) and glutathione reductase (GR) (Mittler, 2000). Glutathione peroxidase (GPX) and glutathione *S*-transferase (GST) also detoxify H_2O_2, lipid hydoperoxides and reactive aldehydes by conjugating them with GSH (Anjum et al. 2012). Likewise, plant cells eliminate MG via glyoxalase (Gly) pathway by which Gly I and Gly II enzymes act coordinately with GSH (Kaur et al. 2014). However, it has been reported that Cd-overloads deplete cytosolic GSH which impinges GSH/GSSG redox potential, causing enhanced Cd sensitivity in several plant species (Flores-Cáceres et al. 2013; Sobrino-Plata et al. 2014).

Hydrogen sulfide (H_2S) has traditionally been thought of a phytotoxin but recently, it has emerged as an important gasotransmitter alongside with nitric oxide and carbon dioxide with multiple functions in animals and plants (Lisjak et al. 2013). H_2S, at tiny doses, can enhance plant growth and development, leading to a sharp increase in global food supplies and plentiful stock for biofuel production (Dooley et al. 2013). Despite enormous prospects of H_2S in sustainable plant agriculture, few studies have focused on H_2S biology in plant systems compared with those in animals. Moreover, mechanisms regulating H_2S functions in plant responses to heavy metal stress, especially in a major crop like rice, are still elusive.

Thus, in the current study, we have examined the regulatory role of H_2S in rice tolerance to Cd stress. Towards this objective, a combined pharmacological, physiological, histochemical and biochemical approaches was employed to identify H_2S-controlled Cd stress tolerance mechanisms in economically important rice plants under Cd stress.

MATERIALS AND METHODS

Plant materials, growth conditions and treatments

Rice (*Oryza sativa* L. cv. BRRI dhan52) seeds were surface sterilized with 1% (v/v) sodium hypochlorite solution for 20 min, washed with distilled water and imbibed for 24 h. The seeds were sown on plastic nets floating on distilled water in 250 mL plastic beakers and kept in the dark at 28 ± 2°C for germination. After 48 h, uniformly germinated seeds were transferred to a growth chamber and grown in a commercial hydroponics solution (Hyponex, Japan) diluted according to the manufacturer's instructions. The nutrient solution consisted of 8% N, 6.43% P, 20.94% K, 11.8% Ca, 3.08% Mg, 0.07% B, 0.24% Fe, 0.03% Mn, 0.0014% Mo, 0.008% Zn, and 0.003% Cu. The seedlings were grown under controlled conditions (photon density: 100 $\mu mol\ m^{-2}\ s^{-1}$, temperature: 26 ± 2°C, RH: 65–70%). Each plastic beaker contained about 50 rice seedlings. The nutrient solution (pH 5.5) was renewed every four days. Uniformly grown 13-d-old seedlings were exposed to nutrient solution supplemented with 250, 500 and 1000 μM of $CdCl_2$ in the presence and absence of H_2S donor sodium hydrosulfide (NaHS, 100 μM). These environmentally relevant $CdCl_2$ concentrations were selected based on previous reports (Naser et al. 2009). To verify the role of H_2S released from NaHS, we applied H_2S scavenger hypotaurine (HT, 200 μM) (Lai et al. 2014) with 500 μM $CdCl_2$ and 100 μM NaHS. Based on literature (Sun et al. 2013, Ali et al. 2014) and our preliminary experiments with a range of NaHS concentrations (25,

50, 100, 150 and 200 μM), we observed that 100 μM NaHS was optimally effective in alleviating Cd-induced toxic symptoms. We also observed that 100 μM NaHS was not effective in alleviating the toxic effects of high dose of $CdCl_2$ (1000 μM $CdCl_2$). Therefore, our experimental design consisted of eight treatments as follows: (i) Control, (ii) 100 μM NaHS (H_2S), (iii) 100 μM NaHS + 200 μM hypotaurine (H_2S+HT), (iv) 250 μM $CdCl_2$ (Cd1), (v) 100 μM NaHS + 250 μM $CdCl_2$ (H_2S+Cd1), (vi) 500 μM $CdCl_2$ (Cd2), (vii) 100 μM NaHS + 500 μM $CdCl_2$ (H_2S+Cd2) and (viii) 100 μM NaHS + 200 μM hypotaurine + 500 μM $CdCl_2$ (H_2S+HT+Cd2). The seedlings were harvested after 3 days of Cd stress treatments to investigate the H_2S-induced mechanisms modulating physiological and biochemical responses. Each treatment was replicated three times under the same experimental conditions.

Determination of physiological and biochemical parameters

Physiological and biochemical parameters such as relative water content (RWC), chlorophyll (Chl) content, proline (Pro) content, lipid peroxidation product malondialdehyde (MDA) content, hydrogen peroxide (H_2O_2) content, histochemical detections of superoxide and hydrogen peroxide, lipoxygenase (LOX) activity, non-enzymatic antioxidants such as ascorbic acid (AsA) and glutathione (GSH), enzymatic antioxidants such as superoxide dismutase (SOD), catalase (CAT), glutathione peroxidase (GPX), glutathione *S*-transferase (GST), ascorbate peroxidase (APX), monodehydroascorbate reductase (MDHAR), dehydroascorbate reductase (DHAR), glutathione reductase (GR), glyoxalase (Gly) enzymes, such as Gly I and Gly II, and trehalose content were measured following standard protocols. Carotenoids content, mineral content, H_2S content, Cd content and histochemical detection of lipid peroxidation by Schiff's test and loss of membrane integrity by Evan's blue test were carried our using standard methods.

Statistical analysis

Data obtained from all the experiments were subjected to one-way analysis of variance (ANOVA) and different letters indicate significant differences between treatments at $p < 0.05$, according to Duncan's multiple range test (DMRT) using IRRISTAT version 3 (International Rice Research Institute, Biometrics Unit, Manila, Philippines). Data represented in the table and figures are means ± standard deviations (SD) of three replicates for each treatment.

RESULTS AND DISCUSSION

Results revealed that Cd exposure resulted in growth inhibition and biomass reduction, which is correlated with the increased uptake of cadmium and depletion of the photosynthetic pigments, leaf water contents, essential minerals, water-soluble proteins and enzymatic and non-enzymatic antioxidants. Excessive cadmium also potentiated its toxicity by inducing oxidative stress, as evidenced by increased levels of $O_2^{\cdot-}$, H_2O_2, MG, and MDA. However, elevating the level of endogenous H_2S improved physiological and biochemical attributes, which was clearly observed in the growth and phenotypes of H_2S-treated rice plants under cadmium stress. H_2S reduced cadmium-induced oxidative stress, particularly by enhancing redox status (AsA/DHA and GSH/GSSG) and the activities of ROS and methylglyoxal MG detoxifying enzymes. Notably, H_2S maintained Cd and mineral homeostases in roots and leaves of Cd-stressed plants. By contrast, adding H_2S scavenger hypotaurine abolished the beneficial effect of H_2S, further strengthening the clear role of H_2S in alleviating cadmium toxicity in rice plants. Collectively, our findings provide an insight into H_2S-induced protective mechanisms of rice plants exposed to Cd stress, thus proposing H_2S as a potential candidate for managing toxicity of Cd, and perhaps other heavy metals, in rice and other crops.

CONCLUSION

Thus, from the aforementioned findings, we can conclude that H_2S could be considered as significant components of plant tolerance mechanisms under Cd stress condition. H_2S provided evident supports to rice seedlings in alleviating toxic effects induced by Cd. Notably, the positive effects was reasonably reflected in the phenotypes of rice plants under such stress, which could be attributed to Cd^{2+} and mineral (Ca, Mg, Fe, Zn, Mn) homeostases, redox re-establishments (AsA/DHA and GSH/GSSG), and reduction of oxidative stress (reduced ROS, MDA and MG accumulation) by regulating antioxidant defense and MG detoxification systems. Finally, our experimental data proposed that the modulation of the level of endogenous H_2S by genetic engineering or external supply could be an important strategy to attain better growth and development of rice, and perhaps other crops, under the heavy metal environmental stresses that would surely augment the sustainable agriculture in this persistently changing environment.

REFERENCES

1. Ali, B., Gill, R.A., Yang, S., Gill, M.B., Ali, S., Rafiq, M.T. and Zhou, W. (2014) Hydrogen sulfide alleviates cadmium-induced morpho-physiological and ultrastructural changes in *Brassica napus*. *Ecotoxicol. Environ. Saf.* 110: 197–207.
2. Anjum, N.A., Ahmad, I., Mohmood, I., Pacheco, M., Duarte, A.C., Pereira, E. et al. (2012) Modulation of glutathione and its related enzymes in plants' responses to toxic metals and metalloids - a review. *Environ. Exp. Bot.* 75: 307–324.
3. Bolan, N.S., Makino, T., Kunhikrishnan, A., Kim, P.J., Ishikawa, S., Murakami, M. et al. (2013) Chapter Four – Cadmium Contamination and Its Risk Management in Rice Ecosystems. *Adv. Agron.* 119: 183–273.
4. Dooley, F.D., Nair, S.P. and Ward, P.D. (2013) Increased growth and germination success in plants following hydrogen sulfide administration. *PLoS ONE* 8: e62048.
5. Flores-Cáceres, M.L., Hattab, S., Boussetta, H., Banni, M. and Hernández, L.E. (2015) Specific mechanisms of tolerance to copper and cadmium are compromised by a limited concentration of glutathione in alfalfa plants. *Plant Sci.* 233: 165–173.
6. Gallegoa, S.M., Pena, L.B., Barcia, R.A., Azpilicueta, C.E., Iannone, M.F., Rosales E.P. et al. (2012) Unravelling cadmium toxicity and tolerance in plants: insight into regulatory mechanisms. *Environ. Exp. Bot.* 83: 33–46.

7. Gill, S.S. and Tuteja, N. (2011) Cadmium stress tolerance in crop plants. *Plant Signal. Behav.* 6: 215–222.
8. Hossain, M.A., Piyatida, P., da Silva, J.A.T., and Fujita, M. (2012) Molecular mechanism of heavy metal toxicity and tolerance in plants: central role of glutathione in detoxification of reactive oxygen species and methylglyoxal and in heavy metal chelation. *J. Bot.* 2012: 1–37.
9. Kaur, C., Ghosh, A., Pareek, A., Sopory, S.K. and Singla-Pareek, S.L. (2014) Glyoxalases and stress tolerance in plants. *Biochem. Soc.* Transac. 42: 485–490.
10. Khan, M.I.R., Nazir, F., Asgher, M., Per, T.S. and Khan, N.A. (2015) Selenium and sulfur influence ethylene formation and alleviate cadmium-induced oxidative stress by improving proline and glutathione production in wheat. *J. Plant Physiol.* 173, 9–18.
11. Kikuchi, T., Okazaki, M. and Kimura, S.D. (2008) Suppressive effects of magnesium oxide materials on cadmium uptake and accumulation into rice grains. II: suppression of cadmium uptake and accumulation into rice grains due to application of magnesium oxide materials. *J. Hazard. Mater.* 154: 294–299.
12. Lai, D., Mao, Y., Zhou, H., Li, F., Wu, M., Zhang, J. et al. (2014) Endogenous hydrogen sulfide enhances salt tolerance by coupling the reestablishment of redox homeostasis and preventing salt-induced K^+ loss in seedlings of *Medicago sativa. Plant Sci.* 225: 117–129.
13. Lisjak, M., Teklic, T., Wilson, I.D., Whiteman, M. and Hancock, J.T. (2013) Hydrogen sulfide: environmental factor or signaling molecule? *Plant Cell Environ.* 36: 1607–1616.
14. Mittler, R. (2002) *Oxidative stress, antioxidants and stress tolerance.* Trends Plant Sci. 7: 405–*410.*
15. Naser, H.M., Shil, N.C., Mahmud, N.U., Rashid, M.H. and Hossain, K.M. (2009) Lead, cadmium and nickel contents of vegetables grown in industrially polluted and non-polluted areas of Bangladesh. *Bangladesh J. Agril. Res.* 34: 545–554.
16. Naser, H.M., Shil, N.C., Mahmud, N.U., Rashid, M.H. and Hossain, K.M. (2009) Lead, cadmium and nickel contents of vegetables grown in industrially polluted and non-polluted areas of Bangladesh. *Bangladesh J. Agril. Res.* 34: 545–554.
17. Ortega-Villasante, C., Rellán-Álvarez, R., Del Campo, F.F., Carpena-Ruiz, R.O. and Hernández, L.E. (2005) Cellular damage induced by cadmium and mercury in *Medicago sativa. J. Exp. Bot.* 56: 2239–2251.
18. Saito, R., Yamamoto, H., Makino, A., Sugimoto, T. and Miyake, C. (2011) Methylglyoxal functions as Hill oxidant and stimulates the photoreduction of O2 at photosystem I: a symptom of plant diabetes. *Plant Cell Environ.* 34: 1454–1464.
19. Seth, C.S., Remans, T., Keunen, E., Jozefczak, M., Gielen, H., Opdenakker, K. et al. (2012) Phytoextraction of toxic metals: a central role for glutathione. *Plant Cell Environ.* 35: 334–346.
20. Sobrino-Plata, J., Meyssen, D., Cuypers, A., Escobar, C. and Hernández, L.E. (2014) Glutathione is a key antioxidant metabolite to cope with mercury and cadmium stress. *Plant Soil* 377: 369–381.
21. Sun, J., Wang, R., Zhang, X., Yu, Y., Zhao, R., Li, Z. and Chen, S. (2013) Hydrogen sulfide alleviates cadmium toxicity through regulations of cadmium transport across the

plasma and vacuolar membranes in *Populus euphratica* cells. *Plant Physiol. Biochem.* 65: 67–74.

THINNING OF APPLES WITH SEQUENTIAL CHEMICAL THINNING SPRAYS

Terence Robinson, Poliana Francescatto, Leonel Dominguez,

Dept. of Horticulture, Cornell University, Geneva, NY

Andrea Rufato, Embrapa, Vacaria RS, Brazil

Gala apple is a difficult to thin cultivar and often is under thinned under NY State conditions. Our objective was to evaluate the usefulness of using sequential thinning sprays to achieve the optimum crop load and crop value. We compared several thinning timings (bloom, petal fall, 12 mm and 18 mm fruit size) and several chemicals (Benzyl adenine=Maxcel + Carbaryl =Sevin XLR, Promalin a 1:1 mixture of BA and GA4+7, ACC and Ethrel). The bloom timing of spraying offers some advantages in potential fruit size and elimination of biennial bearing while the 18mm timing offers a valuable emergency treatment if previous sprays were ineffective. In these studies the bloom sprays of BA were without carbaryl since carbaryl is not allowed at full bloom due to bee toxicity. At the later timings BA was combined with carbaryl. The study was done in Geneva, NY in 2013 and 2014 using an 18 year old 'Gala' apple trees on M.9 rootstock. In 2013 BA applied at full bloom caused significant thinning but not in 2014 where it induced no thinning. BA+carbaryl caused the most fruit thinning at Petal Fall and at 10mm fruit size. It showed intermediate effectiveness at 18mm fruit size compared to the untreated control trees. Fruit size was greatest when BA+carbaryl was applied at 10mm fruit size, intermediate when applied at Petal Fall or 18mm fruit size and smallest when applied at Full Bloom or with the untreated control trees. Sequential applications of BA at full bloom followed by BA+carbaryl at Petal Fall, 10mm fruit size and 18mm fruit size showed increased thinning with increasing number of sprays. Fruit size also increased with increasing number of sprays; however, total crop value was greatest when only the 2 early sprays were applied (Bloom and Petal Fall) intermediate with the 3 or 4 sequential sprays and lowest when only the bloom spray was applied. When the first sequential spray was started at petal fall fruit thinning and fruit size was greatest when all three sprays (PF + 10mm + 18mm) sprays were applied but crop value was highest and similar when two or 3 sprays were applied. At the 18mm timing we compared BA+carbaryl, BA+ACC, Ethrel+oil or BA+carbaryl+oil. BA+carbaryl did not thin but when 0.125% oil was added as a penetrant it caused significant thinning even though fruit size was large. Ethrel+oil also thinned, as did BA+ACC, however, the later caused over thinning. Fruit size was largest with the over thinned BA+ACC, second for the Ethrel+oil, or the BA+carbaryl+oil and smallest for the BA+carbaryl and the untreated trees. Among the 18mm fruit size treatments the BA+carbaryl+oil had the greatest crop value followed by the Ethre+oil and the BA+carbaryl treatments. The BA+ACC treatment had the lowest crop value since it caused over thinning. We plan to focus future trials on both of these unusual timings of bloom and 18mm fruit size.

FOLIAR-APPLIED STIMPLEX® CROP BIOSTIMULANT INCREASES PHOTOSYNTHESIS IN PECAN SEEDLINGS IRRIGATED WITH SALINE WATER

Richard Heerema, Extension Pecan & Pistachio Specialist, New Mexico State University

Sara Moran, Graduate Student, New Mexico State University

Jason French, Graduate Student/Plant Diagnostic Clinician, New Mexico State Univ.

ABSTRACT

Excessive soil and water salinity can present a considerable limitation to photosynthesis, growth, and productivity of pecan (Carya ilinoinensis) in the arid southwestern US. Ongoing drought has caused irrigation water quality to decline and severely limited farmers' ability to apply adequate water volumes for preventing soil salt accumulation. Consequently, salt injury in southwestern pecan orchards has worsened dramatically in recent years. The objective of our study was to measure the impact of soil- and foliar-applied Stimplex® crop biostimulant on photosynthesis of pecan seedlings subjected to salinity irrigation. Thirty-six 1-year-old 'Riverside' seedlings were potted in fine silica sand in 11.4L plastic containers. Beginning on 8 August 2014 irrigation water salinity and Stimplex® biostimulant treatments and were assigned to the seedlings in a completely randomized 2 x 3 factorial experiment. Eighteen seedlings were irrigated with non-saline water [average electrical conductivity (ECw) 0.8 $dS{\cdot}m^{-1}$] during the experiment and 18 were irrigated with saline water [average ECw 4.3 $dS{\cdot}m^{-1}$; created by addition of 0.93 $g{\cdot}L^{-1}$ sodium chloride and 1.17 $g{\cdot}L^{-1}$ calcium chloride to non-saline water]. For both irrigation treatments, six trees received 5$ml{\cdot}L^{-1}$ Stimplex® in irrigation water (soil-applied), six received 5$ml{\cdot}L^{-1}$ canopy Stimplex® sprays (foliar applied), and six received no Stimplex® (Control). Leaf gas exchange was measured on 15 August 2014 with a portable photosynthesis system equipped with a light source and CO_2 injector. For the Control (no Stimplex®) there were, respectively, 48% and 29% reductions in stomatal conductance (g_s) and photosynthesis due to saline irrigation. Under saline irrigation, g_s of those seedlings receiving foliar Stimplex® was more than double that of those receiving no Stimplex® and foliar Stimplex®-treated seedlings exhibited only an 18% reduction in photosynthesis rate due to saline irrigation. These data indicate that foliar Stimplex® application increases photosynthesis in pecan by reducing stomatal limitations to photosynthesis and the negative effects of saline irrigation on photosynthesis appear to be somewhat lessened. Salinity effects on photosynthesis and g_s for the seedlings receiving soil-applied Stimplex®, however, were very similar to that of the Control. Our study suggests that foliar-applied Stimplex® could be used to maintain pecan photosynthesis over the short-term under saline irrigation and likely benefit tree growth and yield. Further work is needed to optimize application rates, determine duration of the benefits to photosynthesis, and investigate the impacts on nut production in field-grown trees.

INTRODUCTION

Pecan (Carya illinoinensis) is a large deciduous tree species in the family Juglandaceae that is native to the south-central United States and to scattered locations throughout Mexico (Grauke and Thompson, 2008). Historically, commercial production of pecan nuts has been found primarily within the native range of the species and in the humid southeastern US (Grauke, 2015). However, especially since the 1970's there has been a notable increase in pecan plantings in the semiarid southwestern US, including New Mexico, Arizona, and far west Texas (USDA, 2015). Today there are nearly 30,000 hectares of pecan orchards in the southwest region, accounting for over 1/3 of total US pecan production (USDA, 2012; USDA, 2015).

Amongst the crops commonly grown in the southwestern states, pecan trees are one of the most sensitive to soil and water salinity stress and injury (Miyamoto, 2006). Salts in irrigation water and soils present one of the most important limitations to photosynthesis, growth, and productivity of pecan in the arid southwestern US. Research conducted in El Paso County, Texas has shown that vegetative growth of pecan trees is inhibited at soil electrical conductivity of saturated paste extract (ECe) of only 2.5 $dS \cdot m^{-1}$ and branch dieback of mature trees occurs when soil ECe reaches 6-8 $dS \cdot m^{-1}$ (Miyamoto, 2006).

Producers in the southwest have a limited range of options for managing salinity in their orchards, all of which require applying irrigation water ("leaching fraction") in excess of crop water requirements. A range of soil management practices, including ripping, deep chiseling, and soil profile modification (mixing) with an excavator are useful for increasing the effectiveness of salt leaching from some southwestern orchard soils (Miyamoto and Nesbitt, 2011). However, ongoing drought in the region has caused irrigation water quality to decline and severely limited farmers' ability to apply the water volumes necessary to adequately leach salts. Salt injury in southwestern pecan orchards has worsened dramatically in recent years.

Applications of seaweed extracts, including from the brown seaweed Ascophyllum nodosum, have been demonstrated to increase the tolerance of some plants to osmotic stresses due to drought, temperature, or salinity (Khan et al., 2009). For example, Spann and Little (2011) showed that a commercially-available seaweed extract applied as a foliar spray or as a soil drench significantly increased the shoot growth and leaf area of potted drought-stressed orange (Citrus sinensis) trees. Furthermore, in the same study they found that the soil drench seaweed extract treatment significantly increased mid-day stem water potential of drought stressed trees. At this point, the mechanism by which seaweed extracts increase the tolerance of plants to osmotic stresses is not at all clear, but it has been proposed that betains, auxins, cytokinins, abscisic acid, mineral nutrients, vitamins, among other components found in the seaweed extracts could play important roles (Khan et al., 2009). If seaweed extract application could similarly increase the tolerance of pecans to osmotic stresses, particularly osmotic stress caused by irrigation water salinity, perhaps the profitability of pecan production could be increased in many areas across the arid southwest.

The objective of our study was to measure the impact of soil- and foliar-applied Stimplex® crop biostimulant on photosynthesis of pecan seedlings subjected to salinity irrigation.

MATERIALS AND METHODS

Our study was conducted in a greenhouse located at the New Mexico State University Fabian Garcia Horticultural Science Research Center near Las Cruces, NM

(32° 16' 46.11" N, 106° 46' 19.70" W).

Thirty-six 1-year-old 'Riverside' seedlings were potted in fine silica sand in 3 gallon plastic containers. 'Riverside' is one of the most common pecan rootstock cultivars used in the western US. Each of the seedlings was pruned to a single shoot axis. Average height of seedlings from the sand surface in the pot to the tip of the shoot axis was 18.1 cm on 31 July 2014. Regular applications were made of granular N-P-K fertilizers to the sand in the pots and foliar micronutrient sprays in order to maintain vigorous growth throughout the study. Trees were irrigated every other day.

Beginning on 8 August 2014 irrigation water salinity and Stimplex® (Acadian Seaplants Limited, Dartmouth, NS) treatments and were assigned to the potted seedling trees in a completely randomized 2 x 3 factorial experiment. Eighteen seedlings were irrigated only with non-saline well water throughout the experiment and 18 were irrigated with saline water. The electrical conductivity (EC_w; measured with a handheld salinity meter) of the well water averaged 0.8 dS/m throughout the study. The remaining 18 trees were irrigated with saline water. The EC_w of the water for the saline irrigation was increased to a target of 4 dS/m by adding sodium chloride (NaCl) and calcium chloride ($CaCl_2$) to the well water at the rates of 0.93 g/L and 1.17 g/L, respectively. Measured EC_w for the saline water averaged 4.3 dS/m.

In both irrigation salinity treatments, six trees received soil-applied Stimplex®, six received foliar-applied Stimplex®, and six received no Stimplex®. Foliar Stimplex® was applied using a hand sprayer to the dripping point at a concentration of 5ml/L. During foliar application, the pots were covered with plastic to prevent drip into the pots and care was taken to prevent cross-contamination of the other treatments. For the soil-applied Stimplex®, the solution was mixed into the irrigation water, which was then poured into the pots (Fig. 1). Addition of Stimplex® increased the salinity of the irrigation water slightly; average EC_w of the tap water became 1.2 dS/m with 5 ml/L Stimplex® and the EC_w of the saline irrigation water became 4.4 dS/m.

Leaf gas exchange was measured on 15 August 2014 using a LiCor 6400XT portable photosynthesis system equipped with a red/blue 6400-02B LED Light Source and a 6400-01 CO_2 Injector (LI-COR Biosciences, Lincoln, NE; Fig. 2). Leaf chamber photosynthetically active radiation was maintained constant at 800 $\mu mol \cdot m^{-2} \cdot s^{-1}$ and reference CO_2 was maintained constant at 400 $\mu L \cdot L^{-1}$. Gas exchange measurements were made 9-10am on one leaf per seedling, choosing a well-exposed, fully-expanded leaflet. A portable chlorophyll meter (SPAD 502; Konica Minolta Inc., Ramsey, NJ) was used to measure greenness of the leaflet opposite the leaflet used for gas exchange measurement.

Analysis of variance tests for growth and physiological data were conducted using SAS Proc GLM (SAS Institute Inc., Cary, N.C.) with mean separation by least means squares analysis. Statistical significance was defined as $P \leq 0.10$

RESULTS & DISCUSSION

Leaf stomatal conductance and photosynthesis rates of potted 'Riverside' pecan seedlings were negatively affected by saline irrigation water (Figs. 3-4). For the Untreated Control (no Stimplex®) there was a 29% numerical reduction in photosynthesis and a 48% numerical reduction in stomatal conductance due to salinity. The salinity effects on photosynthesis and conductance for the seedlings receiving soil-applied Stimplex® were very similar to that of the Control.

Stomatal conductance and photosynthesis rates and of seedlings receiving foliar Stimplex® treatment were elevated compared to that of the Control and the effects of salinity appear to be somewhat lessened (Fig. 3-4). Stomatal conductance and photosynthesis of foliar Stimplex®-treated seedlings irrigated with non-saline water were, respectively, 31% and 19% higher than that of Control seedlings also irrigated with non-saline water. With saline irrigation water, average stomatal conductance of those receiving foliar Stimplex®-treated was more than double that of those receiving no Stimplex®. Foliar Stimplex®-treated seedlings exhibited an 18% reduction in photosynthesis rate when irrigated with saline water. Higher intercellular CO_2 concentration for the plants receiving foliar-applied Stimplex® (Fig. 5) indicates that the increased photosynthesis rates over the short term can be explained in large part by reductions in the stomatal limitations to photosynthesis.

This study shows very promising results for the use of foliar Stimplex® in maintaining, or even increasing, pecan leaf photosynthesis over the short term under saline irrigation conditions, which will likely reflect in increased tree growth and yield. Further work is needed in pecan to fine tune the application rates, determine the duration of the product benefits on photosynthesis, and investigate the level of benefit for field grown pecan trees subjected to saline irrigation water or soil conditions.

SOURCES CITED

Grauke, L.J., 2015. Hickories- *Carya illinoinesis.* USDA-ARS Pecan Genetics, Somerville, TX. 15 October, 2015. <http://cgru.usda.gov/carya/species/illinoinensis/illinoinensis.htm>

Grauke, L.J. and T.E. Thompson, 2008. *Carya illinoinensis* pecan, pp. 421-425. In: J. Janick and R.E. Paul (Eds.), The Encyclopedia of Fruit and Nuts. CABI, Cambridge, MA, USA.

Khan, W., U.P. Rayirath, S. Subramanian, M. N. Jithes, P. Rayorath, D. M. Hodges, A. T.

Critchley, J. S. Craigie, J. Norrie, and B. Prithiviraj, 2009. Seaweed Extracts as Biostimulants of Plant Growth and Development. Journal of Plant Growth Regulation 28: 386-399.

Miyamoto, S. and M. Nesbitt, 2011. Effectiveness of soil salinity management practices in basin-irrigated pecan orchards. HortTechnology, 21(5):569-576.

Miyamoto, S., T. Riley, G. Gobran and J. Petticrew, 1986. Effects of saline water irrigation on soil salinity, pecan tree growth and nut production. Irrig. Sci. 7:83-95.

Miyamoto, S., 2006. Diagnosis and management of salinity problems in irrigated pecan production. Texas Water Resources Institute. 14 October, 2015. <http://texaseden.org/disaster-resources/wp-content/uploads/2011/05/Diagnosis-and-Management-of-Salinity-Problems-pecan-production.pdf>

Spann, T.M. and H.A. Little, 2011. Applications of a commercial extract of the brown seaweed *Ascophyllum nodosum* increases drought tolerance in container-grown 'Hamlin' sweet orange nursery trees. HortScience 46(4):577–582.

USDA, 2012. Census of Agriculture. National Agricultural Statistics Service, Washington, D.C. 15 October, 2015. <http://www.agcensus.usda.gov/Publications/2012/>

USDA, 2015. Noncitrus Fruits and Nuts. National Agricultural Statistics Service, Washington, D.C. 15 October, 2015. http://usda.mannlib.cornell.edu/MannUsda/viewDocumentInfo.do?documentID=1113

Figures

Figure 1. Stimplex® was mixed with irrigation water and poured into containers of 'Riverside' seedlings.

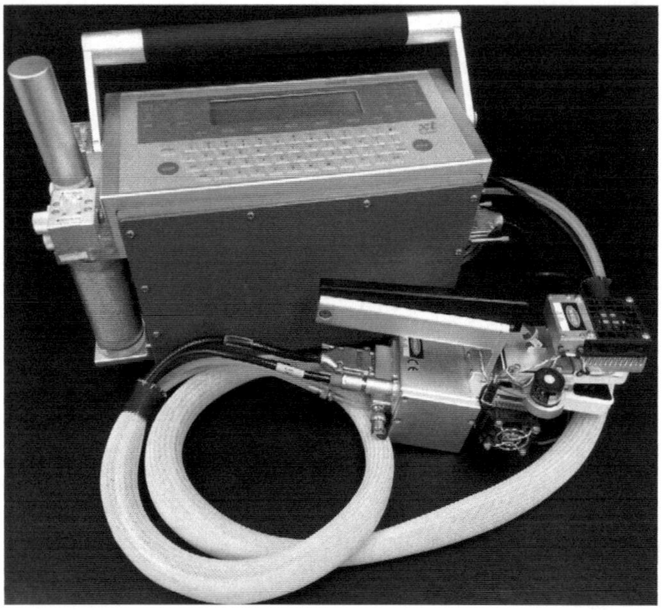

Figure 2. LiCor 6400XT photosynthesis system equipped with light source and CO_2 injector.

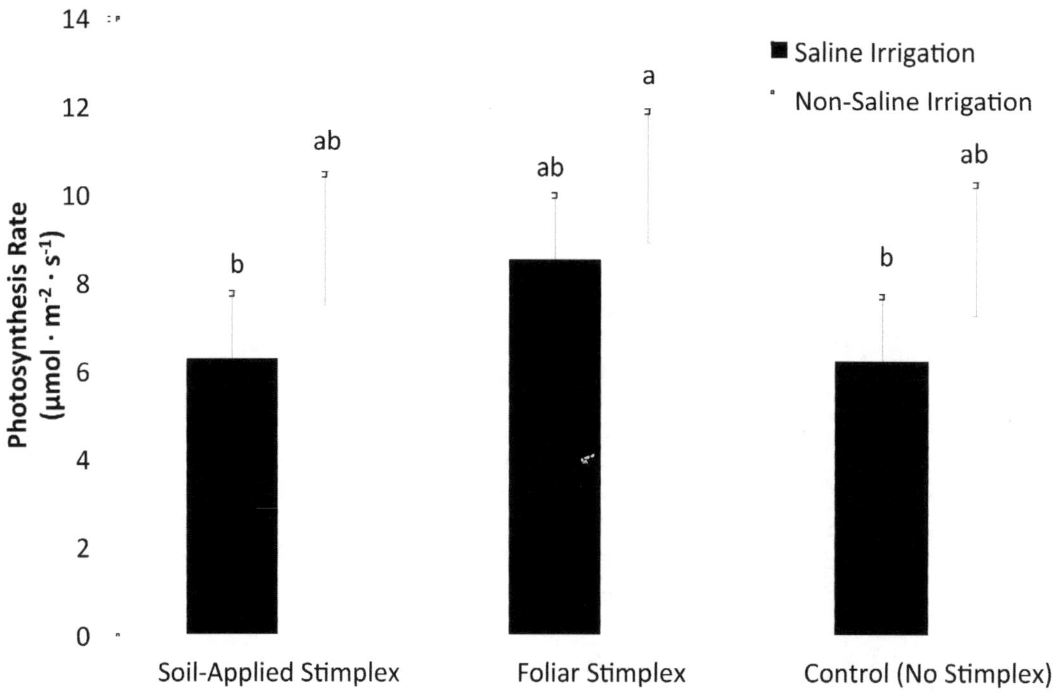

Figure 3. Leaf stomatal conductance to water vapor of 'Riverside' seedlings on 15 August, 2014 following saline or non-saline irrigation treatments and foliar or soil-applied Stimplex® treatments.

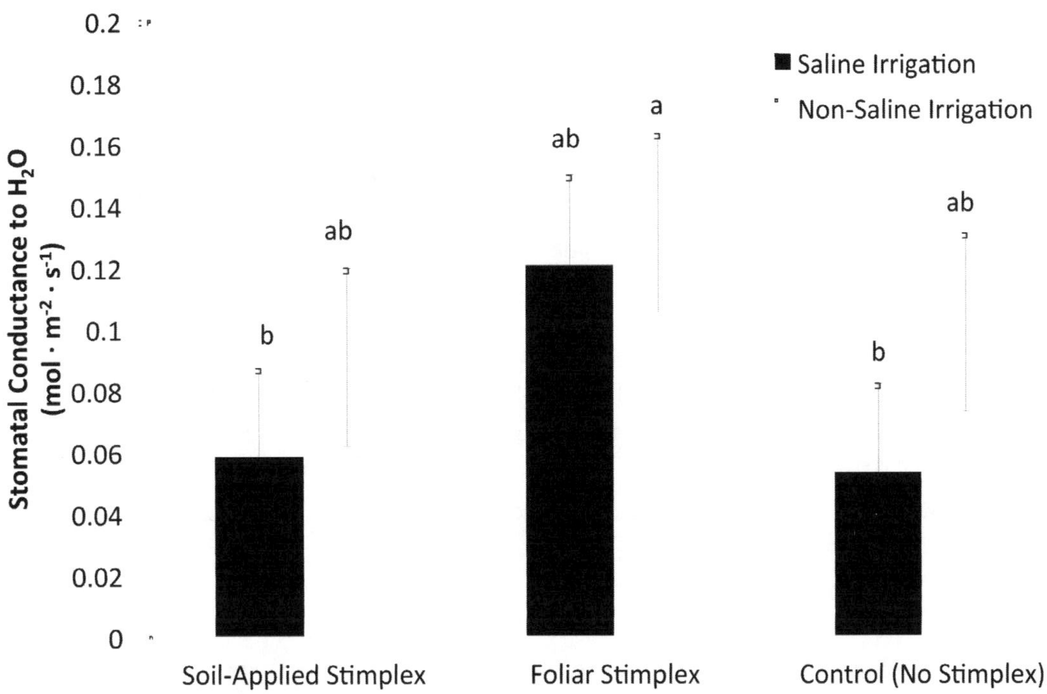

Figure 4. Leaf photosynthesis of 'Riverside' seedlings on 15 August, 2014 following saline or non-saline irrigation treatments and foliar or soil-applied Stimplex® treatments.

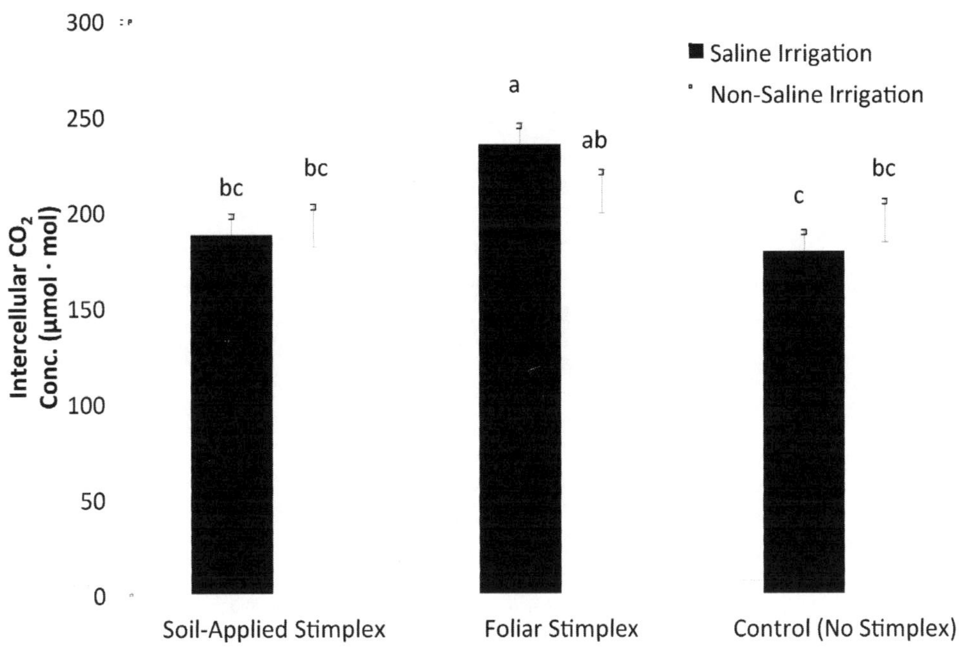

Figure 5. Leaf intercellular carbon dioxide concentration of 'Riverside' seedlings on 15 August, 2014 following saline or non-saline irrigation treatments and foliar or soil-applied Stimplex® treatments.

RFC3 GENE MUTATION LEADS TO REDUCED CELL NUMBER OF LEAVES IN ARABIDOPSIS THALIANA

MO Xu-Dong, Zhang Xiao-Bao, Xiao Lang-Tao, XIA Shi-Tou

Hunan Provincial Key Laboratory of Phytohormones and Growth Development; College of Bioscience and Biotechnology, Hunan Agricultural University, Changsha, China

Systemic acquired resistance (SAR) is a plant immune response induced after a local infection by necrotizing pathogens. Salicylic acid (SA) is an essential signal for the onset of SAR. *replication factor C subunit 3-1* mutant (*rfc3-1*) is a SA hypersensitive mutant which was identified by EMS mutagenesis screen to look for negative regulators of SAR. This mutant has display enhanced and NPR1 independent resistance against *Hyaloperonospora arabidopsidis* Noco2, an oomycete pathogen. *rfc3-1* mutant plants are smaller than wild type plants with narrower leaves and petals. Here to decide whether the smaller plant stature is caused by smaller number of cells, the effects of *RFC3* gene mutation on cell number of leaves was analyzed by flow cytometry method with steps of enzyme digestion of leaves, cell lysis, nuclear collection and dyeing. The results showed that the number of cells in wild type plants was found to be 1.6 times the number for the first and second leaf, and 2.6 times the number for the third and fourth leaf in *rfc3-1* mutant. As a result, the number of cells in *rfc3-1* leaves is sharply reduced. It is further confirmed that the cell proliferation of leaves is affected severely by *RFC3* gene mutation in *Arabidopsis thaliana*.

'DWARF PRATA' BANANA GROWTH AND PRODUCTION UNDER PACLOBUTRAZOL SOIL APPLICATION

Victor M. Maia ,Leonardo C. B. Maia, Danilo S. Barbosa, Lucas B. Ferreira, Ignácio Aspiazú, Rodinei F. Pegoraro

State University of Montes Claros

The aim of this study was to determine the effects of paclobutrazol (PBZ) under growth and production of 'Dwarf-Prata' banana plant clone: Gorutuba. The experimental design was randomized blocks with 5 treatments and 4 repetitions and the parcel was composed of 15 plants and it was evaluated the 3 central plants. The PBZ doses were: 0; 0,5; 1,0; 2,0 e 3,0 g i.a. $plant^{-1}$. The PBZ was applied over the soil in the first production cycle when plants reached 50 cm of height. Were evaluated the pseudostem height (cm) at blooming and the first and second cycle yield. It was adjusted a regression square root model ($\hat{Y}$=207,926-80,6972•√D+25,6447ºD; R^2=0,94) to pseudostem height, where the dose of 2,5 g of i.a. $plant^{-1}$ provided a greater height reduction. Using this PBZ dose the plants reached the height of 144 cm, being 63 cm smaller than the control, which corresponds 30,5% of height reduction. However, doses above 1,3 g i.a. $plant^{-1}$ provided similar plant height reduction. For the first cycle yield, the PBZ doses provided a square response ($\hat{Y}$=13,2862-5,56233*D+1,41303*D2; R^2=0,62). The PBZ provided yield reduction until 2,0 g i.a. $plant^{-1}$ (7,81 ton ha^{-1}). In relation to control, the reduction of yield in dose 2 g i.a. $plant^{-1}$ was 41,2%. The second cycle yield didn't showed significative response to PBZ. This result indicating that the residual effect of PBZ in the soil wasn't sufficient to provide variation at this caracteristic. Support: Fapemig, CNPq and CAPES.

'GRAND NAIN' BANANA GROWTH AND PRODUCTION UNDER PACLOBUTRAZOL SOIL APPLICATION

Victor M. Maia ,Leonardo C. B. Maia, Danilo S. Barbosa,Lucas B. Ferreira, Ignácio Aspiazú, Rodinei F. Pegoraro

State University of Montes Claros

The aim of this study was to determine the effects of paclobutrazol (PBZ) under growth and production of 'Grand Nain' banana. The experimental design was randomized blocks with 5 treatments and 4 repetitions and the parcel was composed of 15 plants and it was evaluated the 3 central plants. The PBZ doses were: 0; 0,5; 1,0; 2,0 e 3,0 g i.a. plant^{-1}. The PBZ was applied over the soil in the first production cycle when plants reached 50 cm height. Were evaluated the pseudostem height (cm) and number of leaves at blooming and yield at first and second cycle. Pseudostem height was adjusted to a quadratic regression model ($\hat{Y}$=229,2339-75,7211•D+14,8667*D2; R^2=0,98), where the dose of 2,5 g i.a. plant^{-1} provided a greater reduction in the plant height. Using this PBZ dose plants height reduced to 132,8 cm, being 96,4 cm smaller than the control, which corresponds to 42% of reduction. However, doses above 1,8 g i.a. plant^{-1} provided similar plant height reduction. The number of leaves didn't showed significative response to PBZ. For the first cycle yield, the PBZ doses provided a linear response ($\hat{Y}$=57,2611-10,5096•D; R^2=0,87). The highest PBZ dose reduced the yield in 31.5 ton ha^{-1} compared to control. It corresponds to 55.1% of yield reduction. The second cycle yield showed a quadratic response to PBZ ($\hat{Y}$=45,2372-14,5430•D+2,8289•D2; R^2=0,99), where the dose of 2,6 g i.a. plant^{-1} promoted greater reduction. This yield reduction corresponds to 41,3%. This result indicating that, in this case, the residual effect of PBZ in the soil was sufficient to reduced the second cycle yield.
Support: Fapemig, CNPq and CAPES.

RHIZOPUS ARRHIZUS KB-2 AS A PRODUCER OF PLANT GROWTH REGULATORY METABOLITES

Yana Evstatieva, Dilyana Nikolova, Aleksandra Ilieva, Valentin Savov

Department of Biotechnology, Faculty of Biology, Sofia University "St. Kliment Ohridski", Sofia, Bulgaria

The genus *Rhizopus* is well recognized filamentous fungi, inhabiting in soil, rotting fruit, vegetables, food, etc. Species of this genus are known as a postharvest deterioration microorganisms causing serious problems in successful storage and marketing of agricultural production. However the species *Rhizopus arrhizus* is technologically important producer of organic acids and hydrolytic enzymes and it is widely applied in biotechnology.

The aim of this study is to estimate the capability of a strain *Rhizopus arrhizus* KB-2 to produce metabolites with plant growth regulatory activity. Morphological characteristics, pH variation and hydrolytic enzyme activity of this strain were defined during the submerged batch cultivation. The presence of some plant regulation metabolites in cultural liquid such as, trans-zeatin, thidiazuron, gibberellic acid, abscisic acid and indole-3-propionic acid was determined by LC-MS analysis.

In order to study the effect of this complex of regulatory metabolites, different dilutions of cell free supernatant was applied in test for germination of *Lactuca sativa* seeds.

THE RHIZOSPHERE, THE PHYLLOSPHERE, AND PLANT GROWTH REGULATION

Andrew J. Krajewski[1] and Shirley A. Krajewski

International Citrus Technologies Pty Ltd, 124, Elizabeth Street, Albany, WA 6330 Australia

[1] corresponding author: citrusdoctor@bigpond.com

Abstract: The plant bridges two vastly different universes, above and below ground, and serves as a conduit for flows of energy, nutrients and information. Many plant physiological processes involve, in some way, microbes. Well known are those colonising the actual surfaces of roots, the rhizoplane, as well as those of the rhizosphere: that biologically active zone of the soil surrounding the plant root, a volume of soil which is chemically, energetically and biologically very different from the surrounding bulk soil. The rhizosphere's contribution to optimal plant growth is out of all proportion to its physical volume. This zone serves as a "broker" between plant and soil through the intimate involvement of a wide range of microbes. They exert both direct and indirect effects on plant growth and development through several complex mechanisms, leading to changes in root architecture, improved shoot and root growth, enhanced nutrient acquisition and tolerance to abiotic stress and protection of plants against pests and pathogens. Less well known are those microbes inhabiting plant leaves and other aboveground surfaces, a zone termed the phyllosphere. Epiphytic colonisation of microbes is the inevitable result of direct exposure of plants to the atmosphere. Certain of these microbes exert effects on plant physiology, including changes in stomatal conductance and transpiration rates, and effects on shoot growth. The paper provides a brief overview of the more important mechanisms by which microbes of the rhizosphere and phyllosphere affect and regulate plant physiology, growth, resistance to stress and defence against pathogens. A better understanding of the ecology of these zones and the processes involved might demonstrate the value of these eco-physiological interactions as cornerstones for robust, productive commercial crop production systems, and might lead to increased appreciation of the role of microbes in soil and plant health.

1.Introduction

We depend upon healthy plants for food, fibre, structural materials and pharmaceuticals and for a wide range of essential ecological services without which we could not exist. The plant acts as the primary agent bridging the above- and below-ground ecosystems, serving as the main conduit for flows of nutrients (including carbon), energy and information (65). Plants connect above-and belowground populations and processes (32).

"Plant health" is a frequently used yet ill-defined term. In this paper, "a plant is healthy, or normal, when it can carry out its physiological functions to the best of its genetic potential" (1; 22).

During photosynthesis radiant energy is absorbed and transformed into the energy of chemical bonds. For every gram atomic weight of carbon taken up through photosynthesis 479 kJ of potential energy is obtained. The energy content of the resulting phytomass ranges between 15 and 35 kJ /gram organic dry matter (most commonly 19-20 kJ g^{-1} (54). The more carbon a substance contains the higher its energy content, and the most important plant substances with very high carbon content are lipids, lignin, and isoprenoids and their derivatives. Consequently, plant parts of highest energy content include resinous leaves, and lignified and suberised organs (54). The capture of radiant energy and its transformation to chemical energy through photosynthesis is the underlying mechanism for production of root exudates (RE: see later) which plant roots secrete into the soil from their roots either by passive processes such as diffusion, through lysis of root cells, or by active secretion (19; 70).

Plant roots serve as conduits for photosynthate from leaves to soil. Plants release into the soil surrounding their roots ca. 12-40% of carbohydrates (CHO) produced by photosynthesis. The so-called rhizodeposition of this vast carbon source makes plants the major driving force of the soil food web (63).

Heterotrophs are those organisms living off fixed energy stored in organic molecules. Plant and animal litter and soil organic matter are the energy resources driving most microbial growth, and the ability of this litter (i.e. its "quality") to provide energy upon microbial decomposition, depends on the type (and number) of chemical bonds present (especially carbon-carbon bonds, but also carbon-hydrogen bonds); the quantity of energy released through their degradation; the size and three-dimensional complexity of the constituent molecules; and their nutrient content (8).

Bacteria require more mineral nutrients in relation to their carbon requirements than either fungi or protozoa, and are thus more likely to immobilise mineral nutrients. However, the activities of fungi, protozoa, nematodes and micro-arthropods generally result in greater release of these nutrients into the soil solution (81).

Microbially-mediated decomposition of organic material usually results in accumulation of stable, chemically complex substances, collectively called humus. When present in large quantity, humus dominates the chemical and biological dynamics of soils. Humus exists in several fractions or forms: humin, humic acid and fulvic acid (8). These substances affect many aspects of plant growth and soil properties (reviewed in 44), typically at relatively low concentrations of ca. 0.001 to 0.0001% (51).

Through interaction with soil microbes, humic substances also serve as energy sources, yielding up to 21kJ/gram of energy available for use by plants and microbes, until further biodegradation takes place (44).

Plants are not just passive recipients of nutrients: information from the environment affects such subterranean allocations as root proliferation; formation of symbioses with mycorrhizal fungi and N-fixing bacteria; and production of defensive compounds. The reciprocal exchange between plants and microbes of signals through transduction pathways may be governed by "rhizosphere control points" coordinating flow and exchange of information between the two (32). Associations of this kind are termed symbiosis, the term first proposed by de Bary in 1879. One form of symbiosis, mutualism, is a special ecological and evolutionary strategy in which two or more dissimilar organisms share their physiological functions between them, rather than being separated (76).

Plants live in an ever-changing, unpredictable environment. Being sessile, they are exposed to many potential stressors. Physiological flexibility is a crucial attribute of plants in coping with abiotic and biotic stresses, and regulation of gene expression is a key element in their adaptation to varying stresses. So-called microRNAs (i.e. small non-coding RNA molecules comprising about 22 nucleotides, with functions in post-transcriptional regulation of gene expression) reportedly play a pivotal role as biological regulators of these plant responses, but also in regulating non-pathogenic plant-microbe interactions, such as rhizobia- and mycorrhiza-triggered symbioses. They do so by either targeting nutritional pathways such as phosphate signalling, or by controlling the microbial colonisation through the "fine-tuning" of transcription factors (9).

Plant-epiphyte cross-talk or communication webs, mediated by an array of chemical substances, may be established as a signal network on the "thin but deep black box" of both the phylloplane and the rhizoplane (reviewed in 35).

In this paper, a general overview is given of the role of microbes in plant growth regulation, with special reference to the rhizosphere. Issues discussed include an overview of organisms involved; the mechanisms by which they promote plant growth; a few examples of specific effects; and examples of stressful conditions under which microbes improve plant fitness.

2. Above-ground: microbial inhabitation of the phyllosphere

The most obvious parts observed are aerial parts: leaves, branches, stems and trunks. The above-ground part of plants, or phyllosphere, is a microhabitat colonised by a wide range of epiphytic inhabitants (85). These congregate on the surfaces of aerial plant parts, and are often nourished by excretions of the latter. They comprise communities of bacteria, fungi, yeasts, algae and protozoa, in very high numbers (51; 85).

Epiphytic inhabitation of microbes occurs due to direct and inevitable exposure of plants to the atmosphere. This phylloplane zone is colonised by microbes undetectable using traditional culture-based methods. Bacteria on leaves are localised at the base of trichomes, in stomata, and along the leaf veins. Phyllosphere microbes are exposed to many changes in their surroundings and survive rapid fluctuations in a hostile environment (53).

Whilst microbial species can be isolated from within plant tissues, many more are recovered from the surfaces of healthy plants (55; 85). When living on leaves, fungal and bacterial colonists of the phylloplane exist as solitary cells, in colonies or in biofilms (85).

Bacteria are by far the commonest leaf colonisers, occurring in numbers averaging 10^6 -10^7 and up to 10^8 cells per cm^2 of leaf surface (55). Per gram of wheat seeds, bacteria numbered from 46,000 to 25 million and fungi 4000 to over 7000. The number of "epiphytes" per gram of green tissue of corn, oats, clover, lucerne (alfalfa) and garden grass varied between 1.5 and 99 million depending upon plant age, soil, climate (51) and geographical location.

Leaves represent an enormous surface: the estimated terrestrial leaf surface suitable for colonisation by microbial epiphytes (mostly bacteria) may be ca. 6.4×10^8 km^2. Clearly, this group of microbes are sufficiently numerous to play a part in the behaviour of the plants they inhabit, and also in many biological processes of global importance (55).

Leaf surfaces are "ecologically relevant landscapes" to microbes encountering or colonising them. Leaf surface topography directly affects availability of suitable microhabitat for colonisation and changes over time at cellular level due to biotic and abiotic processes. These changes significantly affect phylloplane ecology by, for example, altering habitat quality and acceptability by colonising organisms (55; 62). Microclimatic features (e.g. temperature, humidity, wind-speed) are also modulated by leaf surface topography. Rougher leaf surfaces have thicker boundary layers of air than do smooth surfaces. In addition, changes in surface roughness may influence where on the leaf surface various compounds of importance to the microbes are released by cells beneath; distribution of secondary plant compounds in epicuticular lipids; and release of plant-produced volatiles (62). The leaf surface environment provides microbes with nutrients, pigments, terpenes, alkaloids, flavonoids, and presence of volatile plant compounds (35; 85).

Leaf surface roughness affects spectral reflectance from the plant surface. Fluxes in energy whether chemical, in terms of compounds secreted by leaves and utilised by microbes; or physical, in terms of changes in radiant energy due to reflectance, affect the dynamics of microbe-plant interactions on- and within the plant canopy (62).

The phyllosphere is an ecological niche of great significance to both our environment and to agriculture. It is the location for important interactions between microbial inhabitants that may affect fitness of natural plant populations and the quality and productivity of agricultural crops (89).

Mechanisms of plant growth regulation in the phyllosphere

Epiphytic microbes directly affect plant growth by providing nutrients (e.g. nitrogen [N]) and by production of plant growth regulants (PGR: e.g. IAA), as well as indirectly, through inhibition of growth of pathogenic organisms and occupation of niches and competition for resources by non-pathogenic microbes. Bacteria isolated from leaves of *Pinus sylvestris* showed significant ability to produce IAA, produce siderophores, solubilize phosphorus and biologically fix N. Germination assays showed increased root length (by 20-78%); increased hypocotyl length, increased germination rate (by 11 -77%), and increased the vigour index compared with the controls (10; 85).

How these bacteria affect overall physiological processes in plants is still not fully understood. Mechanisms seem to be complex: for example, in *Arabidopsis* plants infected by the phytopathogen *Pseudomonas syringae,* pv *tomato,* stomatal closure is regulated through salicylic acid which closes stomata through an innate immune response. However this pathogen responds by secreting a virulence factor (coronatine) which then causes the closed stomata to reopen (53).

Bacteria (e.g. *Pseudomonas* spp.) can increase leaf wettability by producing compounds with surfactant properties, resulting in enhanced solubilization and diffusion of leaf-produced substrates used by these microbes. This is significant because many pseudomonads colonise the phylloplane and many of the strains produce IAA. IAA production by the phylloplane bacterium *Pantoea agglomerans* was associated with increased fitness of plants during periods of drought stress. Also, there is circumstantial evidence that the synthesis of cytokinin by *Methylobacterium* promotes plant growth, since foliar applications of this microbe increased yield of soybean. In turn, this interaction might trigger release of methanol from plant cell walls and hence provide nutrients to the bacterium (55).

Pink-pigmented, facultatively methylotrophic bacteria of the genus *Methylobacterium* are persistent colonizers of the phylloplane, and are mostly transmitted through seeds. Inoculation of rice plants with a strain of *Methylobacterium* significantly promoted seed germination, increased plant height, plant biomass, number of tillers and grain yield, and significantly reduced the incidence of leaf sheath blight (*Rhizoctonia solani*). Treated rice plants showed increased presence of pathogen-related proteins, phenolics, and enzyme activities (58).

Some phyllosphere bacterial stains (e.g. *Pseudomonas*-, *Pantoea*-, and *Agrobacterium* spp.) have been reported to increase plant growth and nutrient uptake of grasses, cereals and legumes (25). Phylloplane isolates from *Bacillus megaterium* and *Pseudomonas aeruginosa* applied in-field significantly increased plant growth and pod yield of peanut (49). Certain fungal endophytes of leaves may deter herbivores, protect against pathogens and increase drought tolerance (89).

Epiphytic bacteria (*Pseudomonas syringae* and *Erwinia* spp.) possessing ice-nucleating activity ("ice+ bacteria" or ice-nucleation-active bacteria) increase severity of freezing damage to plants by reducing their ability to supercool and hence avoid formation of damaging ice (54; 55; 88).

Finally, phyllosphere-based interactions determine the extent to which human pathogens are able to colonise and survive on plant tissues. This is of great importance in food safety compliance issues where these plant tissues are consumed as fresh salad, fruit and vegetable produce (89).

In summary, the phylloplane clearly plays a number of important roles in terms of plant health and fitness. It is a zone in which radiant energy is intercepted, gases are exchanged, and transport of minerals and other nutrients is driven through the soil-plant-atmosphere continuum. This zone serves as a point of first contact between the above-ground parts of plants and a wide range of phytopathogens and pests. It also provides an extensive surface for the uptake of foliar-applied nutrients and other substances. Many of the microbes inhabiting leaves and other aboveground plant parts play important roles in the regulation of plant growth through a range of direct and indirect mechanisms.

3. Below-ground: the rhizosphere and microbial inhabitation

Life on earth is sustained by the rhizosphere. This relatively small volume of soil concentrates much of Earth's biodiversity of both prokaryotes and Eukaryotes. However, there exists generally a lack of understanding of the rhizosphere as well as of the close connections between the physics, chemistry and biology of this critically important volume of soil (40).

The rhizosphere consists of three zones; the endorhizosphere, comprising portions of root cortex and endodermis in which microbes (and cations) occupy the apoplast; the rhizoplane, the medial zone directly adjacent to roots, including root epidermis and mucilage; and the outermost zone of the ectorhizosphere, which extends from the rhizoplane out into the bulk soil. Although relatively poorly defined in terms of size and shape, the rhizosphere consists of a gradient in chemical, biological and physical properties changing both radially- and along the length of roots (61). The rhizosphere can extend outward by 1 cm or more from the rhizoplane, depending on plant type and soil properties (70). The hyphae of some mycorrhizal fungi extend by up to 10-12 cm from the root surface (33).

All nutrients that plants absorb from soil have to traverse a zone of intense interaction between plant roots, microbes and animals. In 1904, Hiltner called this zone the rhizosphere (13). However, Hiltner's original concept now includes that soil surrounding the living root in which physical, chemical and biological properties have been changed by root growth and activity (71).

In this soil, species richness and abundance are remarkable (40). The diversity of bacteria may range between 6000 taxa and 38,000 taxa in a single gram of soil (compared with ca. 60 in one ml of seawater). As many as 4×10^6 taxa may inhabit a tonne of soil and the number of culturable cells per gram of soil may exceed 10^9, which probably only represents 1-10% of the actual total (63). The total microbial biomass of one hectare of the surface layer of fertile soils amounts to literally tonnes (see table 1: page 35). It is said that the numbers of soil organisms present "are as difficult to understand as the National Debt" (60).

Soil from old Douglas fir stands (Oregon, USA) contained on a dry weight basis about 4,100 kg/ha of fungal mycelium, 5,400 kg of mycorrhizal rootlets, and 40,300 kg/ha of woody roots. Coarse and fine roots made up only 25% of total plant biomass but consumed 40% of the annual, assimilated carbon. Only 3% of total biomass was in fine roots, yet that small percentage accounted for the most intense physiological activity (60).

Even when plants are externally supplied with nutrients, their use depends on the physical, chemical, and especially biological conditions in the rhizosphere. This layer of soil, just a few millimetres thick, is intimately and continuously affected by roots' metabolic processes, creating a zone of intense activity markedly different to the surrounding bulk soil. The rhizosphere's contribution to sustainable soil fertility and thus optimal plant growth is disproportionate to its physical volume. This living interface between roots and soil (63) serves as a "broker" between plants and the soil they inhabit (69).

Soil microbes do more than cycle and deliver mineral nutrients to plants. They exert strong positive and negative effects on plant growth and development, depending upon the species and external conditions. Microbes in soil constitute the most important and indispensable link in the nutrition of plants. They create the conditions necessary for plant growth and influence this growth through their metabolic products (51).

Roots greatly influence the accumulation of soil microbes and their activities. Large differences exist between numbers of bacteria found in rhizosphere soil compared with bulk soil (51; 63) and also in such soil properties as pH (differences of 2 to 3 pH units); reductase activity (by factor >100); increase in root-exuded soil constituents such as sugars (two to three-fold), amino acids (10-fold) and citrate (60-fold); increased phosphatase activity (two to 8 fold); depletion of extractable P and K (5- to 7-fold); accumulation of calcium (up to 10-fold, resulting in its precipitation); and accumulation of NaCl (6-fold) in saline soils planted to various crops (69).

Lysenko in 1955 considered that the rhizosphere microflora acted as "the digestive organ of (the) plants" (51), and the soil microbial biomass was described as "the eye of the needle through which all nutrients must pass" (91). Interactions in the rhizosphere between plants and soil biota jointly mediate processes such as soil aggregation, nutrient cycling, hydrology and carbon storage. However, in agroecosystems, the significance of rhizosphere contribution has been diminished by inputs of synthetic fertilizer, herbicide, fungicide, and other interventions such as tillage and irrigation (15; 23).

Amongst those microbes in the rhizosphere are root-inhabiting plant growth promoting rhizobacteria (PGPR) which aggressively colonise plant roots. They reportedly promote plant growth through at least one of the following mechanisms: suppression of plant disease; improved nutrient acquisition (e.g. of N, P, S, Fe); or plant growth regulant (PGR) production (including IAA, cytokinins, gibberellins, abscisic acid, and inhibitors of ethylene production: 3; 46; 71). These mechanisms are discussed later.

Nutrient acquisition by plants in the rhizosphere involves complex, highly ordered interactions between plant roots, soil microbes and soil animals. The concept was put forward of "rhizosphere control points" to emphasise the importance of information exchanges between plants and microbes through gene expression patterns, with resulting morphological and physiological changes for both parties involved (66).

Plants play a key role in selecting and enriching the types of microbes that inhabit the rhizosphere through the production of root exudates (RE). The microbial community develops depending upon the nature and concentration of organic RE constituents and their ability to utilise these as energy sources. There is a continuum of bacterial presence through the rhizosphere, on the rhizoplane and within the plant tissues, with a wide range of associations that are highly significant to plants and to plant production (13; 53; 71). Microbes can affect root exudation through effects on permeability of root cells; root metabolism; and the absorption or excretion of certain constituents of RE by microbes (70).

The exudation in favour of soil microbes of resources, which otherwise may have found use in additional plant growth, supports these activities of organisms soil in performing processes directly beneficial to plant growth (13).

3.1. Root exudation

Near the roots of growing plants a special zone is formed in which more favourable conditions exist for both microbes and roots. As early as the 18th century, roots were known to secrete or exude substances affecting soil properties and thus fertility (51).

The soil system is different from that above ground: it is not driven by abundant solar energy and is therefore energy-limited. Root exudation has been described as a "moveable feast" (42). These carbon-rich exudates support many microbial interactions in the rhizosphere, and are especially important to the range of free-living soil organisms known to be strongly carbon-limited (13; 32; 61).

Root exudates, plant litter and animal waste deposited above ground are the primary sources of energy for the belowground community. Root exudates include secretions actively released from roots into soil; and diffusates passively released due to either osmotic differences between soil solution and root cells, or as lysates from autolysis of epidermal and cortical cells and dead fine roots. The process is called rhizodeposition. Root exudates are chemically complex: inorganic constituents include a range of inorganic ions (e.g. OH^-, HCO_3^-, H^+) and gases (CO_2, ethylene and H_2).

Organic compounds released include low- or high molecular weight species. The former are the most diverse (organic acids; amino acids; proteins; sugars and CHO; vitamins; phenolics and other secondary metabolites generally more easily used by microbes) and have a wider array of known or potential functions. However, the high molecular mass organic substances (e.g. mucilage, cellulose) make up the majority of C released in this way. The range of these substances, along with plant species, edaphic and climatic conditions, both influence and are influenced by the soil microbial community. This affects nutrient acquisition (e.g. P and Fe); allelopathy (not here discussed); chemical signalling (chemotaxis) to attract symbiotic partners (e.g. rhizobia and legumes); and promotion of colonisation of root surfaces by beneficial microbes (19; 61; 70; 81; 86).

Given our increasing interest in carbon, a wide range of CHO occur in soils in such quantity that they may represent 4 to 7% of total organic carbon, with annual flux (in stands of *Picea alba*) of up to 70kg C /ha/year. Sugars serve as chemo-attractants of rhizobacteria and CHO such as gluconic- and alginic acids play a role on sequestration of such heavy metals as Cr-VI (86).

Roots release 10- 250 mg C per gram of root produced (i.e. ca. 10-40% of total carbon fixed by photosynthesis), both as organic- (e.g. organic acids) and inorganic forms (e.g. HCO_3^-), although organic C- compounds are the most varied and most greatly influence the biological, chemical and physical processes in the rhizosphere. Factors affecting rhizodeposition include plant type; climatic conditions; herbivory; nutrient status; and soil properties (61).

Differences in amount and composition of RE are likely to most affect microbial community composition because microbial species differ in their ability to compete for- and metabolise different carbon sources. Plant species, however, differ in their response to specific microbes (42). Mycorrhizal plants transfer more photoassimilates to their roots than non-mycorrhizal plants, perhaps as the result of higher demand for carbon by these fungi for their growth and development (59).

Although plant growth depends on light, moisture and available energy, nutrient availability in the rhizosphere may be directly influenced by plant roots, through such mechanisms as induced pH changes through secretion of organic acids, and increased chelation trace elements (e.g. Fe, Mn and Zn) by organic- and amino acids. However, because of simultaneous release of energy resources by roots, soil bacteria also take up soil nutrients. A "microbial loop" was proposed to exist, triggered by release of RE from plants which increases bacterial growth in the rhizosphere. This increased growth sequesters nutrients within bacterial biomass, and nutrients remain "locked up" until the bacteria are consumed by predators such as protozoa and nematodes and excreted as waste products which are then available (now in mineralised form) for uptake by plants. Of significance is the biomass of the protozoa in soil (see table 1); their high production rates (10-12 times their standing crop per year); and minimum generation times (i.e. 2 -4 hours). The grazing-induced changes in soil microbial function affect key ecological processes, because soil bacteria occupy some of the most important control points for nutrient cycling and hence plant growth. As an example of this, N-fixing, nitrifying and denitrifying soil bacteria regulate the nitrogen cycle. Protozoan grazing often stimulates nitrifying bacteria through predation of their faster-growing bacterial competitors, in turn leading to high nitrate concentrations in leachates of rhizosphere soil (11; 12; 13; 32).

A high proportion of RE is reportedly used and released as CO_2 in a very short time, with only a small proportion incorporated into microbial biomass. The perceived low efficiency of this conversion (relative to the theoretical maximum and when compared with other carbon sources in the soil) may be due to the co-existence in the rhizosphere of biological N fixation. This process requires large quantities of energy, especially where of an associative nature with free-living rather than symbiotic N fixers. At least 16 ATP molecules are consumed to convert one N_2 to two NH_3 molecules, in addition to the other processes required for associative N fixation. Hence, where a high proportion of RE is used by diazotrophic microbes in the rhizosphere, this might account for the perceived low assimilation efficiency of these substances (18).

Organic carbon exuded from roots is not the only resource required by rhizosphere microbes. Transpiration-driven movement of water through soil carries nutrient and non-nutrient salts from bulk soil through that of the rhizosphere through the gradient that drives the SPAC. This flux of soluble salts can result in concentrations of these substances many times that of bulk soil. Conversely, nutrient ion uptake by roots drives nutrient movement by diffusion and creates zones of depletion of such nutrients as ammonium, nitrate and phosphate (32; 36).

In addition to stimulating microbial growth, RE establish and maintain contact between roots and soils, lubricate growing roots and protect them from desiccation (63).

3.2. Soil microbes and plant growth

The best known key microbe-plant relationships include biological N fixation in the legume-rhizobia symbiosis (e.g. 10; 14; 20); mycorrhizal fungal associations (6; 14; 33;45; 59; 60; 73); and plant growth promoting rhizobacteria (PGPR) (2; 3; 17; 21; 24; 25; 26; 28; 38; 46). A brief overview follows of their overall significance processes relating to plant growth promotion.

3.2.1. Nitrogen fixers

Geochemical weathering of parent rock produces negligible amounts of N, hence N supply is a critical factor in developing plant communities and ecosystems. Most soil N comes from biological N fixation (BNF) by prokaryotic organisms either free-living; loosely associated with plant roots; or in mutualistic symbiotic associations with plant roots (14; 42). These interactions affect competitive advantage and hence ability of plants to exploit different ecological niches. Plants possessing these associations usually rely on their symbionts even more when in competition with other plant species (42). Many of these associations are of great significance to agriculture, especially where there is no or limited reliance on inputs of synthetic N fertilizer.

Biological nitrogen fixation is accomplished by bacteria and Archaea. Together these diazotrophs may contribute ca. 180 x 10^6 tonnes of elemental N per annum globally, of which 80% comes from symbiotic relationships (e.g. obligate symbionts like *Rhizobium* in legumes and *Frankia* in non-leguminous plants) and the remainder from non-symbiotic, free-living-, associative-, or endophytic systems (e.g. cyanobacteria, *Azospirillum*, *Azotobacter* and others: 71).

Symbiotic N fixers abound in many forms worldwide. The best known of these belong to the genus *Rhizobium* and are bacteria that, through a highly complex process (not here discussed), enter the root hairs of legumes and induce them to form small nodules which enclose the bacteria, and shield them from oxygen while simultaneously granting access by gaseous N_2 in the soil atmosphere to this "living factory". The host plants also supply the energy (as sugars and other photosynthates) required by the rhizobia to fuel the process of N fixation (60).

Several groups of true bacteria are also free-living N fixers, including *Azospirillum*, a bacterium which lives in constant association with mycorrhizal fungi in Douglas-fir forests. *In vitro* experiments suggest that *Azospirillum* is difficult to culture in the absence of the fungus, and it seems unable to fix N without its fungal partner. This raises doubt as to the number of so-called "free-living" bacteria that are actually "free". Plants forming N-fixing nodules with symbiotic bacteria also depend on mycorrhizal fungi for their nourishment, with this fixed N feeding into plant tissues and returned into soil by leaf fall and root exudation of nitrogenous compounds from leaves and roots (60).

Cyanobacteria (comprising the blue-green bacteria and green- or chloroxybacteria) occur in both free-living and symbiotic forms. Cyanobacteria perform the dual roles of photosynthesis and N-fixation. Free-living N-fixers may contribute annually up to one thousand million tonnes of fixed N worldwide (60).

Apparent symbioses are those most likely to develop host-specificity. In legume-rhizobia relationships, for example, specificity is determined by plasmids which contain the genes for both nodulation and N-fixation, and can be exchanged among strains (36).

Non-symbiotic N fixers such as *Azotobacter, Azospirillum, Bacillus* and *Pseudomonas* are also of great agronomic significance, and interactions with plants lead to a wide range of beneficial effects in many plants (71).

3.2.2. Mycorrhizae

All truffles, mushrooms and many other fungi produce hyphal networks that grow between plant rootlets to form a shared absorptive organ known as a mycorrhiza (83). "Mycorrhiza" is a general term for the symbiotic relationship between a soil fungus and a plant root. Unlike rhizobia and their legume partners, mycorrhizal associations are relatively non-selective and ubiquitous, occurring in ca. 80% of angiosperms and in all gymnosperms. This form of symbiosis may have arisen ca. 450 million years ago, and hence greatly predate the associations between rhizobia and legumes, which might explain why this form of symbiosis is now so common through the plant kingdom. Although parasitic and neutral associations exist, most of these associations benefit both host plant and colonising fungi (61).

Host plants benefit through assistance in acquiring water, phosphorus and such micronutrients as zinc and copper through provision of an increased contact area between plant, fungus and soil. The fungi benefit from transfer of a wide range of carbon sources from the plants. Mycorrhizae differ in their dependence on plants for carbon. The ectomycorrhizae housed within the intercellular spaces but not within the cells of root tissues and the endomycorrhizae that grow into the root cortical cells and enter the cells can demand up to 20-40% of all the carbon that the host plant fixes through photosynthesis. Endomycorrhizae are wholly dependent on the host plant for their carbon requirements (61).

Mycorrhizae colonise roots and develop an external mycelium which serves as a bridge connecting roots with the surrounding soil microhabitats. Thus, the mycorrhizal symbiosis, in linking biotic with geochemical components of the ecosystem, effectively contributes to nutrient acquisition and supply. The arbuscular mycorrhizal (AM) symbiosis in particular plays a direct role in nutrient cycling rates and distribution patterns which affect root exudation in plants, and physical properties of the soil surrounding the roots, in turn affecting the rhizospheric microbial communities in this so-called mycorrhizosphere. Conversely soil microbes can affect AM formation and function, with the so-called mycorrhiza helper bacteria able to stimulate mycelial growth of these fungi or enhance mycorrhizal formation (82). These relationships may greatly benefit plants.

Most plant species form symbiotic relationships with mycorrhizal fungi and nearly two-thirds of all land plant species have arbuscular mycorrhizal (AM) symbiosis, which usually leads to improvement of performance of both fungal and plant partners. Arbuscular mycorrhizal fungi cannot be host-specific because only ca. 200 fungal "species" (the meaning of which is hard to define in these obligately asexual organisms) have been described. However, these can colonise about 200,000 potential host plant species. Host plants, furthermore, often acquire several different AM partners over time, with certain of these proving to be most suitable. Study of these specific AM fungi is complicated by the fact that only about 70 species being successfully cultured in pots, a fact which itself casts doubt on the applicability of these "easy to culture" species to the field situation (42).

An intricate chemical dialogue initiates mycorrhizal associations with plants, the study under controlled conditions (i.e. in vitro) of which is complicated by the fact that endomycorrhizae are obligate symbionts and cannot be cultured independently of their host plants. More is known about the ectomycorrhizae because they can be grown independently of their host. In both cases, however, it is suggested that the fungus can sense either the presence of volatiles (e.g. CO_2) or chemicals in plant root exudates, which initiate hyphal growth and branching. Conversely, chemical signals from mycorrhizae which are perceived by plants have long been suspected. These *Myc*-factors, only recently chemically characterised, are similar in structure to the rhizobial *nod* factors, and it is unclear whether the fungus produces these signals all the time or only when it senses that the plant is near (61).

AM fungi absorb nutrients from soil, including N, P, K, S, Ca, Mg, Fe, Mn, Cu and Zn, and translocate them to the roots of associated plants. The efficiency with which mycorrhizal plants take up P is 3.1 to 4.7 times higher than with non-mycorrhizal plants due to the extension of external fungal hyphae into soil zones with un-depleted P. The external hyphae of some AM fungi may spread up to 10-12 cm from the root surface (33).

Enabling associated plants to maintain a more balanced supply of nutrients is another advantage that AM associations can confer where nutrient shortages exist in soils, often as the result of the "lock-up" of these in insoluble forms, including precipitates, or due to negative interaction with other elements in soil solution. Nutrients most commonly affected are P, as well as Ca, Fe, Mn and Cu. The effects of P-induced micronutrient deficiencies are well documented, especially Zn and Cu, also, high P levels are known to inhibit mycorrhizal associations, resulting in reduced uptake of certain mineral nutrients, and hence a state of nutritional "imbalance" (33). The precise nature of nutrient "balance" or imbalance is not clearly defined.

Not all AM associations benefit the plants. There exists a "mutualism-parasitism continuum" : some AM are more beneficial to their hosts than others. Unequivocal evidence exists that some soil microbes can decisively influence plant competitive interactions and that interactions amongst microbes can determine the outcome of plant competition (42).

In order to acquire nutrients for the plant, both ecto- and endomycorrhizae extend hyphae centimetres into the soil, resulting in a 10-fold increase in effective root surface area and a 2- to 3-fold increase of P and other nutrients per unit root length compared with no mycorrhizal plants. In addition to high density of the hyphae, nutrient uptake is facilitated by their narrow diameter (ie. 200 µm) which enables them to access very narrow soil micropores not accessible to thicker root structures (61). The dense, intertwined network of hyphae form a "common mycorrhizal network" in which hyphae of mycorrhizae infecting two or plants are connected, and through which plants not sharing the same space have been shown to share nutrients. Arbuscular mycorrhizal hyphae interconnect root systems of plants, and evidence exists of the movement (although in relatively small quantity) of isotopes of C, N and P between plants through AM hyphae (33).

The range of interactions also includes improvement of soil aggregation (see section 5.1., page 26) and structural stability, resulting in improvement in such soil properties as aeration and internal drainage. This in turn leads to improved growth of plants, which influences plant diversity, ecosystem function and stability (61), and agricultural productivity.

Since mycorrhizal fungi differ from other symbionts in that they are not transmitted vertically from host parents to offspring, successful mycorrhizal colonisation is established by complex communication between fungi and plant. The initial recognition step probably indicates a metabolic switch from the asymbiotic- to either a pre-symbiotic or an infective state. Furthermore, early recognition steps do not activate an unregulated cascade of events leading to symbiosis development. Rather, each new morphogenetic change in both partners requires a successful sequence of biochemical communication (73).

Mycorrhizal fungi have potential to both influence and respond to rhizobacteria. Impacts of rhizobacteria on mycorrhizal fungi include changes in root-fungus signalling; recognition and receptivity; and effects on fungal germination and growth (63).

Finally, there is increasing evidence that the composition of AM fungi communities is determined more by soil conditions than by the composition of plant communities (42). This demonstrates the interdependence of soil physical, chemical and biological properties.

3.2.3. Bacteria

As unicells, bacteria possess the capacity to rapidly respond to change, since their ribosomes attach themselves to mRNA before transcription from DNA is complete, thus enabling rapid metabolic responses to changes in their environment. Rhizobacteria are therefore probably genetically predisposed to rapid colonisation and exploitation of the rhizosphere. Also, with their very high metabolic capacity, they play important roles in many functions affecting plants and soils, including nutrient cycling; plant growth promotion; symbioses and pathogen suppression (63).

Plant growth promoting rhizobacteria (PGPR) have been extensively reviewed (e.g. 3; 38;46;71) and were first defined in 1978 as organisms that, after being inoculated onto seeds, successfully colonise plant roots and enhance plant growth. These are the most widespread plant co-habitants, comprising a taxonomically diverse group including non-pathogenic, gram-negative bacteria (e.g. *Azospirillum, Burkholdia, Enterobacter* and *Pseudomonas*); gram-positive genera (e.g. *Bacillus, Paenibacillus* and *Streptomyces*); and certain Archaea (76). These microbes directly and indirectly promote plant growth through a range of mechanisms, discussed later (25; 61; 75).

These bacteria produce other compounds that promote plant growth. Organic acids and vitamins play important roles in soil. Aliphatic (e.g. acetic-, citric-, fumaric-, malic-) organic acids occur in concentrations up to 6000 μM in some soils. Phosphate solubilizing bacteria (e.g. *Arthrobacter, Bacillus, Delftia, Phyllobacterium, Rhodococcus*) which increase P-uptake by plants produce such aliphatic organic acids as citric, lactic and propionic acids, and citric, oxalic and malic acids are believed especially important in rhizosphere ecology and pedogenesis (86). Phosphate solubilizing bacteria, azotobacters (*A. chroococcum* and *A. vinelandi*) and rhizobia are all significant producers of vitamins including β-carotene (the precursor for vitamin A, or retinol); ascorbic acid (Vit. C); tocopherol (Vit. E); and a range of vitamins B (e.g. biotin, riboflavin, niacin, thiamine, pantothenic acid). Although information on quantities of vitamins found in soil is scarce, they apparently play an important role in soil biochemical soil processes (86).

From the microbial point of view, evolution of strategies enhancing energy transfer to the roots (through root exudates) would be expected to increase fitness of those microbes able to influence gene expression by signalling plants. Most of the specialised rhizobacteria seem capable of producing plant growth substances. About 80% of bacteria isolated from rhizospheres are able to produce auxins, and their widespread ability to produce cytokinins is well known. This capacity for production of both plant growth substances and other signal molecules suggests that rhizosphere bacteria play important roles in manipulating root and plant growth (11).

Chemical signalling allows single bacterial cells to act in coordination with an entire bacterial population or community in response to changes in the rhizosphere environment. Such coordination within a microbial community is important to ecological relationships such as competition or cooperation that, although of an adaptive nature, would have a lower probability of success if undertaken solo. Many biotic interactions involve chemical signalling in response to environmental clues, and include pathogenesis; antibiosis; release of extracellular enzymes; motility; biofilm formation and initiation of symbiosis (36).

Microbial colonisation is not uniform over the root system and may cover 15-40% of the total plant root surface. Common sites for bacterial attachment and colonisation are at epidermal cell junctions, root hairs, axial grooves, cap cells and sites of emerging lateral roots. Root exudates serve as both food and chemoattractants to microbes, which attach and may form microcolonies. These microcolonies can eventually grow into biofilms, comprising various arrangements of multiple layers of bacteria encased in an exopolymeric matrix. Bacterial cells within biofilms are physiologically distinct from the same cells in so-called "planktonic form" grown in dispersed culture. The effectiveness of PGPR at promoting growth commonly depends upon their density: once a critical microbial density is reached, the biofilms begin to act in unison through a process termed quorum sensing, coordinating the release of compounds promoting plant growth through the direct and indirect mechanisms previously discussed. It is clear that biofilms can- and do form on biotic surfaces during plant host-microbe interactions and most plant-microbe interactions rely on (close) physical interaction between bacteria and plant tissues such as root hairs (61; 67; 74).

For example, *Azospirillum brasiliense* and related species are motile, heterotrophic proteobacteria interacting with the roots of such cereals as wheat and maize, promoting growth of the plants. Although *A. braziliense* is a free-living N-fixer, the bacterium colonises root elongation zones and root hairs, forming dense biofilms, where its ability to promote plant growth seems related to stimulation of root formation, rather than to N fixation. In this and many other instances, the properties of biofilms to increase adherence of structural components is so important that it has been suggested that the process of biofilm formation *per se* drives- or significantly affects the dynamics of plant-microbe interactions (67). As the result of this, these assemblages appear to show elevated tolerance to stresses such as other competing microbes as well as increased ability to promote plant growth (74).

Root-root and root-microbe communications probably occur continuously in the biologically active rhizosphere. These intercellular signals enable a bacterial population to control the expression of genes in response to cell density. Many root-colonising bacteria produce N-acyl homoserine lactone (AHL) as a signal to coordinate their behaviour in local rhizosphere populations, and such processes as nodulation and successful infection by pathogenic strains may depend upon AHL-mediated "quorum sensing" regulation (87). A common breakdown product of AHL is homoserine lactone, which induces a strong increase in transpiration rate of certain plants. This may benefit soil microbes where soil moisture (i.e. the soil solution) carries mineral nutrients through mass flow towards the roots (32).

Plants are able to affect density-dependent behaviours of rhizobacteria by secretion of quorum-sensing mimics or interference compounds. Soil microbes have many signalling mechanisms that affect plants in turn. Bacteria can sense and respond to PGR from plants and release PGR analogs, resulting in either positive or negative effects on plant growth (36).

There are many examples of PGPR improving plant growth through multiple mechanisms. For example, PGPR *Bacillus* spp. strains positive for P-solubilization, siderophore production and IAA production were used to test effects of seed inoculation of scarlet runner bean plants. Inoculation significantly increased net leaf photosynthesis; transpiration rate; water use efficiency and leaf chlorophyll content. Increased superoxide dismutase and peroxidase activity was also observed during periods of intense photosynthesis. These two antioxidants tend to detoxify H_2O_2 accumulated during periods of intense photosynthesis, and efficient removal of reactive oxygen intermediates from chloroplasts is critical, since H_2O_2 concentrations as low as 10 μM may decrease photosynthesis by 50%. Finally, treatment significantly increased bean yield, and soluble protein- and total reducing carbohydrate content of shelled bean seed (79).

3.3. Interactions and Synergies: mycorrhizal fungi and PGPR

As already stated, many PGPR can stimulate plant growth through direct or indirect interactions with plant roots. In addition most plants are colonised by mycorrhizal fungi and their presence also generally stimulates plant growth. However, these two groups of "beneficials" have mainly been studied separately, with relatively few studies on possible synergistic effects in promoting plant growth. Both ecto- and endomycorrhizal fungi can interact with these bacteria, with these interactions occurring in the zone of soil surrounding the roots and fungal hyphae, commonly known as the "mycorrhizosphere" (6).

Under natural conditions, mycorrhizal fungi are surrounded by a complex bacterial community that interact with these symbiotic fungi. Of two basic types (saprophytic and symbiotic), these gram –positive bacteria can be neutral, harmful or beneficial. Beneficial forms, having the ability to enhance root colonisation or promote hyphal growth of both ecto- and endomycorrhizas, are called mycorrhiza helper bacteria (MHB) (6; 39).

Many such bacteria are able to attach to roots, AM and other fungal hyphae, and provide direct or indirect benefits to host plants through multiple mechanisms (76). As well as directly influencing the mycorrhizal relationship or plant growth itself, specific bacteria together with mycorrhizal fungi may create indirect synergism supporting enhanced plant growth through improving nutrient acquisition, enhancement of root branching or inhibition of plant pathogenic fungi (6).

Interactions between rhizobacteria and AM fungi might be mediated by surface contact or through soluble factors. In studies on attachment between these partners, a two-step mechanism might operate. During the first stage, weak binding occurs, governed perhaps by general physicochemical factors such as electrostatic attraction, whereas in the second stage, the more stable binding might result from mechanisms involving production of cellulosic fibrils or other extracellular polymers. The significance of rhizobacterial attachment for mycorrhizal functioning was not clear (6).

Bacteria can live within AM fungi, with both gram-positive and negative bacteria on the surface of AM spores but only gram-negative ones inside the spores themselves, where they appear to live intracellularly associated with lipid and protein bodies. Chitinolytic bacteria isolated from AM spores could play a role in spore germination (59). Mycorrhizal helper bacteria stimulate mycelial growth and enhance mycorrhiza formation, and perhaps affect the receptivity of the root (24). These microbes can produce compounds that increase root cell permeability and rates of root exudation, in turn stimulating mycorrhizal fungal mycelia in the rhizosphere or facilitate root penetration (mycorrhization) by the fungus (6; 39; 72).

Benefits appear to flow in both directions. Various AM fungi increase population density of rhizobacteria. For example, *Glomus* spp. colonisation increases population density of the following bacteria or bacterial groups: *Azotobacter, Azospirillum brasiliense, Actinomycetes, Streptomyces* spp., ammonia oxidisers, Mn reducers and N fixers (59). However, *in vitro* studies showed increased spore germination (doubled) and increased hyphal length (by between 87% and 520%) of *Glomus* species by six tested strains of *Pseudomonas* bacteria, an effect attributed to IAA production by these MHB (39).

Nitrogen-fixing bacteria improve bioavailability to plants of N, a capability enhanced when plants are also colonised by AM fungi. For N_2-fixing rhizobia the mycorrhizal and root-nodule symbioses are synergistic both to in terms of infection rate and impact on mineral nutrition and plant growth. Uptake of other essential micronutrients from soil by AM fungal hyphae may also result in general plant growth improvement and also indirect effects on the N fixation system. Although enhanced N fixation (section 4.1., page 18) is apparently mediated by generalised stimulation of nutrition of the host plant, more specific effects probably occur at the level or the root or nodule. Arbuscular mycorrhizal fungi may also interact with both symbiotic and free-living N-fixing bacteria (6).

Bacteria may also assist the AM symbiosis through increasing the bioavailability of phosphorus. This is especially significant in soils with low P bioavailability. Free-living P-solubilizing bacteria release phosphate ions from sparingly soluble inorganic and organic compounds in soil, thereby increasing P for uptake through the hyphae to be passed on to the plant. Inorganic P is usually released by bacterial secretion of various organic acids, whilst organic P is released by the action of bacterial secretion of phosphatase enzymes. Phosphate-solubilizing bacteria (PSB) are known to interact with AM fungi: these bacteria promote mycorrhizal establishment whereas the mycorrhizal symbiont increases the size of the bacterial population. Hence, AM fungi and PSB interact to make use of P sources otherwise unavailable to plants, with resulting beneficial effects on growth and fitness of plants (6).

Co-inoculation of *Bradyrhizobium* (another N-fixing bacterium relatively slow-growing compared with *Rhizobium*) with certain PGPR can positively affect N fixation by enhancing both root nodule number and dry mass of nodules (71).

Rhizobium's effects on legumes are indirect, and include increased nodulation and increased yield through provision of N to such crops as chickpea, pigeonpea, soybean, groundnut and moongbean. However, rhizobia can also improve plant P nutrition by mobilising both inorganic- and organic P from soils. Hence, use of both rhizobium and PSB (which also produce PGR) appears to exert a synergistic effect on plant performance (71).

Several fluorescent pseudomonads increase rates of ectomycorrhiza formation on *Eucalyptus diversicolor* seedlings by up to 300%, leading to significant increase in seedling biomass. Available data indicate that these interactions are very complex, but release by MHB of volatile compounds may play a role in fungal growth (45).

Increasing emphasis is laid on the use of two or more microbes to maximise beneficial plant growth responses. A microbial-based approach using compatible strains of AM fungi and Rhizobium has been used in rehabilitation of desertified ecosytems and wastelands, and with saline-, drought-prone and calcareous soils, resulting in improved plant performance (72).

Understanding the mechanisms governing the interaction between various microbes in the plant rhizosphere is the key to improving both the level and reliability of plant growth stimulation (26), and forms the basis of a number of powerful remediation and crop production technologies, discussed later.

4. Effect of soil microbes on plant growth promotion

There are many examples of plant growth promotion by soil microbes (see 2; 3; 37; 46; 71 for reviews). However, a detailed account of the specific beneficial effects of various microbes and their activities on a range of plants will not be covered here, and the interested reader is referred to the following references: 6; 17; 20; 21; 23; 25; 28; 30; 31; 34; 44; 45; 46; 48; 51; 52; 53; 57; 64; 72; 75; 77; 79; 82. However, selected examples of effects will, where appropriate, be given for each of the following mechanisms through which soil microbes promote plant growth.

4.1. Biological nitrogen fixation (BNF).

In 1888 Hellriegel and Wilfarth proved that improved yield of crops grown in rotation with legumes was due to conversion of atmospheric di-nitrogen (N_2 gas) into ammonia, a form useable by plants, by *Rhizobacterium leguminosarum*, a bacterium housed within nodules on the roots of leguminous plants. This led in turn to Hiltner's postulation in 1904 of the rhizosphere concept. Since then, this interaction has been shown to consist of a "dialogue" between rhizobium and plant, initiated by a chemical (flavonoid) signal released when plants are under N deficiency. This signal induces nodulation (*nod*) genes in the rhizobia, leading to encoding of a *nod* factor: enzymes necessary to produce a lipochitooligosaccharide chemical response. The *nod* factors initiate a cascade of developmental processes in the host plant root, which facilitates colonisation by the bacteria and formation of the nodules in which these bacteria are eventually housed (61). The process of nodule formation is comprehensively detailed in (10). Biological nitrogen fixation is also carried out by symbiotic genera such as *Azorhizobium, Bradyrhizobium*, and *Frankia* (71).

Phenolic substances in root exudates serve as chemoattractant signals to microbes and affect *nod* gene induction and inhibition in rhizobia (19). When the white lupin (*Lupinus albus*) plant is subjected to N deficiency stress, its roots specifically release the isoflavone

2'-hydroxylupalbigenin, both a *nod*-gene inducer as well as a chemo-attractant for *Rhizobium lupini*, the symbiotic nodulation bacterium for this plant. The plant responds to this signal at the root surface: the root hairs curl and branch, and mitosis actively occurs in the cortical tissues underneath the root hairs to allow nodulation (35).

Nodules formed on legume roots vary widely in structure and size, but all have internal zones infected by the rhizobium to form symbiosomes, or bacteroids. These include the cells containing nitrogenase, which converts gaseous N_2 from the air in the soil first into ammonium and then into other nitrogenous compounds such as ureides, glutamine and/or asparagine. These are the most important compounds that transport N from the roots via the xylem to the plant's aerial tissues (10).

The *nod* factors are generally species specific, and account for the host specificity observed between rhizobia and their plant partners. McNear (2013) noted that this specificity is astonishing considering that one gram of soil contains $>10^9$ bacteria probably from 10, 000 species, of which a very small proportion have developed the ability to symbiotically cohabit with plants. This ability is believed to have evolved around 60 million years ago (61).

Concerning BNF, Brazilian sugar cane growers use N-fixing organisms which directly improve plant growth and productivity. The quantities of applied N fertilizers are typically 65 kg N/ha/annum to produce yield of about 80 t/ha fresh cane weight. This contrasts markedly with application of ca. 150-300 kg N/ha/annum by producers in the USA, Australia, India, Mexico and Colombia. Australian cane production using 200 kg N/ha/annum plus irrigation was 84 tonnes per ha, but at greatly increased capital cost (10). The ecological aspects of high N applications in the context of runoff from cane fields and resulting pollution of ground water and of estuaries and coasts in these areas are also ongoing matters of concern.

In legumes, a major by-product of N fixation is hydrogen gas (H_2) the production of which consumes about 5% of net photosynthesis. In rhizobia possessing the uptake hydrogenase (HUP+) capacity, hydrogen produced by nitrogenase activity is oxidised to yield more energy. However, with HUP- strains, this (un-reacted) gas is evolved into the rhizosphere. Of interest is that pre-treatment of soil with hydrogen can stimulate growth of crops by 10-30%, and in hydrogen treated soils 60% of the reducing power was found to flow to oxygen, and 40% to carbon dioxide, resulting in net chemoautotrophic CO_2 fixation (19).

Other N-fixing, nodule-forming microbes are actinomycetes ("moldlike bacteria"). These form nodules larger than those found on legumes, and which branch repeatedly in a coral-like pattern. These colonise plants such as Acacias, and Casuarinas, and are important N fixers in many habitats (60).

The interaction with plants of non-symbiotic N fixers such as *Azotobacter, Azospirillum, Bacillus* and *Pseudomonas* spp. results in increased yield of annual and perennial grasses, sugarcane, sugar beet, potato, radish, wheat and chickpea; to increased seed germination in maize and mung bean; and increased root length and dry mass in mint (71).

4.2. Solubilization, mineralisation or mobilisation of nutrients

Mere contact of roots with soil is apparently insufficient to secure the full nutrient requirements of plants (4; 11; 14. Microbes serve as the mediators between nutrient sources in soils and plant root systems (11; 14; 51; 69). Their involvement in the cycling of such key plant nutrients as C, N, P, and S are extensively documented (e.g. 14), and will not here be discussed.

Soil microbes are both sinks for nutrients, but also represent sources, since microbes themselves contain significant quantities of nutrients. Uncultivated soils in Western Australia can contain 20-60 kg N/ha "tied up" in microbial biomass. In a 16-year stubble-retention rotation (wheat at the time of sampling) the quantities of "microbial" carbon recorded were 229 kg C/ha in the soil depth 0 to 10 cm; 112 kg C/ha in 10-20 cm; and 69 kg C/ha at 20-30 cm (43). Other reports on quantities of nutrients in soil microbial biomass recorded 108 kg N/ha; 83kg P; 70kg K; and 11kg Ca in the upper 12.5 cm depth of soil (5). Although not directly available to plants, these nutrients occur in appreciable quantities, and in close proximity to living and active plant feeder roots. Given the short generation time and high turnover rate of many of these microbes, the operation of the processes in the soil food web and resulting liberation of "plant-available" nutrients contribute significantly to the health of plants.

The strong top-down regulation of rhizobacteria through nematode and protozoan predation (or "grazing") confers a central role to these organisms in plant root-soil microbe interactions. A strong stimulating effect is commonly observed in nitrifying bacteria when protozoan grazers are present, possibly through predation of their faster-growing bacterial competitors, and subsequent release of nitrates in both liquid culture and rhizosphere soil (32). Estimated rates of N mineralisation due to bacterial-feeding nematodes are 0.0012-0.0058 µg N per nematode per day, mainly as NH_4^+, and N-mineralisation by fungal-feeding nematodes as 0.0018-0.0033 µg N per nematode per day (91). Considering that nematodes number some 1000 per g of soil and up to 150 kg of nematodes per ha (14; 40), amounts of mineralised N released into the rhizosphere, in close proximity to roots, may be considerable.

An active microbial community is fundamental to the supply of P to plants. Microbes themselves constitute a large fraction of the total P in most soils. Hence their significant impact on soil P turnover is not surprising. Microbially-derived nutrients have shorter turnover times than those originating from organic matter and highest rates are usually found in rangeland soils, with annual soil P fluxes of about 40 kg P per hectare. With organic P mineralisation, daily rates (determined by different methods) range from 0.2 to 2.5 mg P per kg of soil (84).

Developments in agricultural sustainability involve rational exploitation of soil microbial activities and use of cheaper, though less bio-available, sources of plant nutrients, such as rock phosphate, which is rendered available by microbiological processes. For example, so-called "bio-available soil P" is that which leaves the solid phase of the soil and enters the soil solution. This is a process in the plant-soil system involving soil microbes; especially PGPR (84).

Phosphate solubilizing bacteria (PSB) include *Bacillus, Pseudomonas, Paenibacillus, Xanthomona* and *Rhizobium*, and *Aspergillus* and *Penicillium* are two fungi that effectively do the same (30; 46; 71; 75). These organisms are able to release bio-available P from insoluble forms of phosphorus in soils through their secretion of organic acids which directly dissolve the phosphates, or by chelating calcium from insoluble tricalcium phosphate (a common form of "fixed" phosphate in soils) with resulting release of P into the soil solution (71).

Bacillus megaterium (strain TRS-4) isolated from the rhizosphere of tea plants (*Camellia sinensis*) was tested on plants in-field and in pots. Inoculation using aqueous suspensions and "bio-formulations" resulted in increased plant height and emergence of new leaves and branches, and increased leaf number and leaf dry mass, with differences evident after 2 months in potted plants and 12 months in-field. *B. megaterium* also increased accumulation of phenolics in plants and solubilization of phosphates *in vitro* and *in vivo*. Application of this bacterium decreased soil P content, enhanced soil phosphatase levels, and increased leaf- and root P content and leaf chlorophyll levels (17).

Some strains of endophytic *Rhizobium* increased solubilization of phosphate fixed in plant-unavailable tricalcium phosphate, by secretion of extracellular organic acids (e.g. citric) as well as phosphatase enzymes, leading to increased growth of rice (20).

Although many soil microbes are able to solubilize P from sparingly soluble organic or inorganic P sources *in vitro*, the contribution of this process to plant nutrition is clouded by the possible re-fixation of solubilized P ions on their way to the root surface. This microbially- solubilized P could however be taken up by the mycorrhizal mycelium, thereby serving a synergistic microbial interaction (82).

Concerning nutrient availability, the scarcity of bio-available iron (Fe) in soils and on plant surfaces "foments a furious competition". Under Fe-limiting conditions, PGPR (*Bacillus, Enterobacter, Pseudomonas* and *Rhodobacter*) produce low molecular mass compounds called siderophores, which are high-affinity Fe-chelating compounds also secreted by other organisms such as fungi and by grasses. The production by microbes of these compounds exerts a direct effect on the availability of Fe for plant growth. Siderophores such as enterobactin are amongst the strongest binders of Fe known (71).

4.3. Production of plant growth regulants (PGR)

Gibberellic acid was the first metabolic product of soil microbes (in this case from the fungus *Gibberella fujikuroi*) to be obtained in a chemically pure form (51).

A wide range of PGR substances (also called phytohormones or plant growth substances) are found in soils. PGR coordinate information from many sources through many points of regulation, from biosynthesis to transduction, and those produced by rhizosphere microbes or provided by inocula in the root vicinity directly affect plant growth and development after being taken up by the roots (48).

Concentration of auxins range widely depending upon soil properties. More auxin is found in soils enriched with additions of organic materials such as manures and composts. In the "A" horizon of so-called "fertile" soils, up to 175 µg "auxins" per kg of dry soil were reported (compared with 40-60 µg per kg in "nonfertile" soils). Auxins certainly exist in concentrations sufficient for the stimulation of plant growth, ranging in different soils between 0.16 to 0.045 µg equivalents of indoleacetic acid (51).

IAA is the main auxin in plants, controlling such properties as cell division and enlargement and responses to light and gravity. Hence bacterial IAA producers interacting with plants have the potential to interfere with such processes through interfering with the plant's native auxin pool. The consequence depends usually on the amount of auxin and on the sensitivity of the plant to changes in auxin concentration. Tumour- and gall-inducing bacteria (e.g. *Agrobacterium tumefasciens*; *Pseudomonas savastanoi*, *Pantoea agglomerans*) were the first plant-associated bacteria in which IAA synthesis pathways were studied (78).

In various PGPR (e.g. *Azospirillum*), production of IAA results in increased root proliferation, in turn resulting in enhanced mineral uptake and root exudation, which in turn stimulates bacterial colonisation and thus amplifies the inoculation effect. IAA biosynthesis alone does not account for the overall plant growth stimulation, so the "additive hypothesis" was advanced to explain *Azospirillum*'s effect on plant growth and yield. Growth promotion was postulated to result from the concerted action of multiple mechanisms, including PGR synthesis, N fixation and others (78).

Bacterial strains differ in their ability to produce auxins, and also perhaps as the result of availability of substrates. Both IAA and IBA may be produced, in varying concentration and with and without the addition of tryptophan as an auxin precursor. Substances produced by bacteria are released continuously, and especially where the bacteria colonise and grow on plant surfaces or within internal tissues. The PGR produced probably improve plant growth through direct effects on metabolic processes. However, since some phosphorus solubilizing bacteria induce both proliferation of lateral roots and their dense coverage with root hairs, the increase in nutrient absorptive capacity probably leads to increased nutrient absorption, which would account for the observed increase in length of plant shoots and roots (75).

Besides being a well-known PGR, IAA is increasingly recognised as a microbial metabolite and signalling molecule. Bacteria use IAA to interact with plants as part of their colonisation strategy, which includes phytostimulation and circumvention of basal plant defensive mechanisms (78).

Treatment of seeds or cuttings with certain non-pathogenic bacteria (e.g. *Agrobacterium, Alcaligenes, Bacillus, Comamonas, Paenibacillus, Pseudomonas, Streptomyces*) induced root formation through production of auxin (28).

In addition to auxins, soil microbes such as *Azospirillum, Azotobacter, Pseudomonas* and *Rhizobium* spp. produce gibberellins, cytokinins and abscisic acid, and such active substances as ACC-deaminase. *Bacillus* spp also produce bioactive volatiles such as 2,3-butanediol and acetoin (3; 38; 71).

Several reports suggest that, under conditions of iron deficiency, a strain of *Pseudomonas aeruginosa* is able to produce salicyclic acid, which has been variously classified as a siderophore (21) as well as a plant hormone (37).

4.4. Enhanced resistance to biotic stresses

Plants commonly experience stresses such as drought, oxidative stress, heavy metals, activities of pathogens and pests (2; 26; 54) and certain management practices (16). In coping with stresses, resources allocated to maintenance of homeostasis, defence or repair and restoration cannot be simultaneously allocated to other physiological and metabolic processes in the affected plants. Stresses often occur in combination, e.g. drought and salinity stress (14; 26; 54). Effects of such combinations are difficult to predict, and plants under stress often pay a premium in terms of their fitness. Management practices acting as stressors include over-application of certain fertilizers; lack of provision of essential trace elements; herbicide application; and repeated foliar sprays of chemicals that affect plant metabolism (e.g. organophosphorus and carbamate insecticides and fungicides). In general relatively little attention is given to the combined effect on plant health and production of a range of stressors acting in concert.

Beside auxins, soils also contain thiamine, riboflavin, biotin, vitamins B6 and B12, inositol, nicotinic acid, folic acid, pantothenic acid and others. Of significance is that concentrations of these are highest in the rhizosphere. Of interest is that thiamine and biotin have been found in soils at depths of greater than 2 m. The microbes are seen as the main factor enriching the soil with these and many other substances (48; 51; 86). Chemosynthetic and root-nodule bacteria (e.g. *Bacterium, Bacillus, Clostridium, Pseudomonas, Rhizobium, Thiobacterium*); fungi (*Aspergillus, Candida, Mucor, Penicillium* spp) and mycorrhizas, and algae (e.g. *Chlorella* and *Hordeum* spp.) are all reported to produce biotic substances capable of eliciting in a wide range of plant growth effects. Plants can obtain vitamins and amino acids from microbes, through the capacity of microbes to transfer these to plants (51).

Plant fitness may also be directly improved through production by fungal endophytes of substances which are toxic or disruptive to plant pathogens and pests. These include antibiotics, with microbial production of single or multiple kinds, as well as polypeptides, aromatic compounds, alkaloids and terpenes, all of which have significant effects at low concentrations. Endophytes also secrete a range of lytic enzymes capable of hydrolysing polymeric compounds including chitin, proteins, cellulose, hemicellulose and DNA. When endophytes colonise their host plants, they use these enzymes to hydrolyse plant cell walls in order to invade plant tissues. These enzymes also directly suppress pathogen activity through degrading cell walls of fungi and oomycetes. Also, microbial activity and degradation of organic matter results in production of plant growth promoting humic substances (29; 44; 48; 86).

Indirect effects enhancing plant resistance to pathogen or pest invasion include the well-documented induction of plant resistance through two basic resistance responses: systemic acquired resistance (SAR), which is induced by pathogen infection, mediated by salicylic acid, and leads to accumulation of pathogenesis-related proteins; and induced systemic resistance (ISR), induced by certain non-pathogenic rhizobacteria, mediated by jasmonic acid and ethylene, and which does not lead to accumulation of pathogenesis-related proteins. The proteins comprise a range of enzymes, some of which act directly to lyse invading cells (e.g. chitinases and glucanases); reinforce cell wall boundaries to resist infection; or induce the "hypersensitive response" of localised cell death (3; 29; 71).

A second indirect effect is where endophytes contribute to host plant defences through control of plant physiological processes, likely through their production of PGR. Improved plant growth reflects increased vigour, and is thus at least potential protection to pathogen challenge. Many studies show that plants infected with endophytes show increased growth, tolerance to drought and to unfavourable soil conditions. As plant defence responses increase demand for energy, resources such as carbon skeletons and "reducing equivalents" are provided through primary metabolic pathways (54), and enhanced plant growth induced by endophytes will enhance indirect protection from pathogens to the benefit of plants (29).

The use of rhizobacteria and their effect on induction of induced systemic resistance (ISR) is well documented, and belowground biota and their interactions can stimulate defence signals from the roots to the aboveground plant parts, either through air-borne or vascular signals (53).

PGPR can increase plant growth indirectly by suppressing plant pathogens through a range of mechanisms including carbon substrate competition; sequestration of Fe^{3+} by siderophores that the pathogen cannot then access; by increasing Mn availability in soil; production of antibiotic substances; and the induction of SAR in plants (63).

Production of siderophores potentially provides a dual mechanism for microbe-mediated growth promotion of plants: enhanced uptake of Fe for the plant and suppression of rhizosphere pathogens unable to utilise the Fe-siderophore complex. However, this mechanism does not appear to function with all strains of rhizobia (20).

It was shown that aerial infection of *Arabidopsis* plants by the pathogen *P. syringae* pv. *tomato* led to the recruitment of the beneficial rhizobacterium *Bacillus subtilis* FB17 in *A. thaliana* root. This was due to increased L-malic acid secretion. The rhizobacterial biofilm formation increased in response to the aerial pathogen, ultimately making the plant more tolerant to infection (53).

Multiple members of the genus *Bacillus* were shown to be capable of modulating stomatal opening and closure yet not when the rhizobacteria are added to the leaves. Also, in the absence of an active competitor during root colonisation, the extensive growth of beneficial rhizobacteria is quite uncommon, indicating the possibility that such soil microbiota associate with plants only at a time of distress. The switch between defence and benefit is crucial since it concerns the cost-effectiveness of allocation of plant resources (where plants are "primed" for protection) and affects plant survival (53).

Low molecular mass organic acids are believed responsible for minimising crop damage by root-knot nematodes (*Meliodogyne incognita*). Production of gluconic acid by rhizosphere bacteria reduces predation by protozoa by increasing encystment or mortality of protozoan predators. Acetic, lactic and citric acid secreted by the microbes were all shown to have antibacterial effects (86).

Fluorescent pseudomonads isolated from the rhizosphere of groundnut showed suppression of the economically important pathogen *Macrophomina phaseolina* (causing charcoal rot of groundnut). Two pseudomonad strains tested showed capacity to produce siderophores, IAA and HCN, and also caused phosphate solubilization under normal growing conditions. Inoculation of groundnut with these strains enhanced seed germination by 15 to 30% with subsequent increase in grain production by 66-77%, compared with treatment with pathogen alone, which decreased yield by 57% (77).

Bacteria are by no means alone in their ability to suppress pathogens and pests. The fungus *Trichoderma* serves as a bio-control agent through a range of mechanisms, suggested to include mycoparasitism; antibiosis; competition for space and nutrients; induction of SAR; inactivation of phytopathogen enzyme systems; and siderophore production, which sequesters iron needed by pathogens (64).

4.5. Ecological interactions suppressing diseases

Ecological effects associated with endophytic colonisation of plant tissues are also, although indirect, a significant mechanism for plant growth promotion. Many different microbes and organisms occupy the soil rhizosphere niche, and compete strongly for space and resources. Fungal endophytes may, by their rapid colonisation, protect plants through out-competing the pathogens. These fungi dominate utilisation of the limited substrates available, even to the point of their exhaustion. Also, the plant responds to invasion by the non-pathogenic endophytes by increasing the deposition of lignin and other cell wall components. As a result, the cell walls become reinforced after endophytic invasion, making it more difficult for pathogens to breach these walls and infest the plant tissues. Since not all endophytic interactions benefit the host plants further study is needed to clarify precise mechanisms involved (29).

A very efficient defence against phytopathogens may be provided by microbes creating protective barriers on root surfaces, with *Pseudomonas* (spp. *fluorescens, putida* and *chlororaphis*), *Bacillus* spp. *cereus* and *subtilis*, and also *Serratia marcescens* being capable of defence against such pathogenic fungi as *Fusarium*, *Rhizoctonia* and *Verticillium*, through a range of host protection mechanisms (76).

In addition to other direct mechanisms, plant growth promotion may also result from suppression of deleterious microflora by the introduced endophyte. Tomato seedlings inoculated with a *Pseudomonas* strain showed increased plant growth accompanied by dense colonisation of the applied strain in internal root tissues and consequent displacement of pathogenic indigenous endophytic pseudomonads. The beneficial effects of bacterial endophytes vary, but appear to operate through mechanisms similar to those in PGPR. However, because of the different habitats they colonise, endophytes offer another tool useful in biological control strategies (34).

Many plants resist soil-borne diseases through an effect called rhizosphere camouflage. Although the rhizosphere of these resistant plants is occupied by high numbers of bacteria, the types of these microbes are more similar to those in the bulk soil than to those associated with the disease-susceptible plants. Thus, having a rhizosphere microbial community similar to that existing in the bulk soil may make the rhizosphere of these plants less identifiable to pathogens. Cultural practices adding large quantities of fresh organic matter may cause the bulk soil community to more closely resemble that of the rhizosphere, assisting rhizosphere camouflage and suppression of root diseases (14).

The issue of so-called "suppressive soils" was first reported in Australia in 1971, and has since been found in many other locations. It concerns the pathogen *Phytophthora cinnamomi*, which affects a host range of over 1000 plants, and devastates such plants as jarrah (*Eucalyptus marginata*), various *Banksia* species, cinnamon (*Cinnnamomum burmannii)* and avocado. The suppressive effect has been attributed to collective efforts of several soil microbes with antagonism toward the fungus, rather than to a single bacterial or fungal isolate. Mechanisms likely include complex interaction between abiotic (e.g. soil pH, clays, nutrient status) and biotic factors (the soil microbial community and interactions with the pathogen and plant roots) although the suppression is seen as primarily biological in origin (14; 47).

Numerous *Bacillus* and *Paenibacillus* strains express activities that suppress pests and pathogens. Specific management strategies are likely to be required to maximise the beneficial effects of these interactions (30). Inoculation with these organisms is increasingly widely seen as providing alternatives to use of synthetic fertilizers and plant growth promoters in sustainable agriculture (26; 38).

5. Effects on soil properties and root architecture.

5.1. Soil properties

Soil physical, chemical and biological properties are interlinked and interdependent, and changes to any one of these will tend to result in changes to the others. Soil microbes, especially in association with plant roots, play a number of extremely important roles in all properties of soil.

Many soil microbes, especially fungi, strongly affect the genesis and diagenesis of sediments, minerals and rocks and thereby influence soil formation. In attacking soil minerals, fungi and bacteria apply two mechanisms: acid attack by secretion of inorganic (e.g. carbonic-, nitric-, sulphuric-) and especially organic (e.g. oxalic-, citric-, acetic-, etc) acids; and mechanical activities that disaggregate or aggregate particles and thus increase or decrease reactive surface area (80). Mineralogical studies of rhizosphere in forest soils showed that inorganic soil particles within these zones are mechanically, chemically and mineralogically affected by activity of soil microbes. Mineral grains abutting root surfaces show evidence of increased fracture, and such biogeochemical weathering in the root zone contributes to soil genesis (58).

Fungal attack on soil minerals results in alteration of carbonates; dissolution and precipitation of metals ions; degradation of silicates; processing of sulphur and phosphorus; biosorption of heavy metals; and degradation of fossil carbon. The many physiological activities of fungi influence geological processes and hence plant nutrition. Also, their manner of growth and morphology affect aggregation of soil particles (80).

Microbially-produced CHO contribute to soil structural properties such as soil particle aggregation, with secretion of such fungal metabolites as extracellular CHO performing important roles in aggregate stability through mineral-organic associations (14; 86). Mycorrhizal fungi themselves secrete exudates such as glomalin, variously described as a glycoprotein (63) or as a collection of iron-containing proteins. Glomalin is an important biotic regulator of water-soluble aggregation in bulk soil and has a mean turnover time in soil of 6 to 40 years (23).

Glomalin production by *Glomus intraradices* was suggested to be "extremely plastic" and appears to respond to such environmental factors as soil pore structure. Several rhizobacteria (e.g. the N-fixer *Paenibacillus* endemic to the wheat rhizosphere and *Rhizobium* sp. in the rhizosphere of sunflower) produce exopolysaccharides which modify soil structure and increase water holding capacity around the root system, leading to increased drought tolerance of plants (23). Since microbes are mainly found in soil aggregates, glomalin can increase microbial population densities (59).

Soil microbes in the rhizosphere further contribute to soil aggregate stability through synthesis of compounds from the decomposition of soil organic matter (14; 44). Also, fungal hyphae (especially those of AMF) contribute to soil stability by enmeshment of aggregates or through cross-linking with them (23; 59; 63).

Roots of higher plants and their associated microbes have coevolved with soils. They collectively play a major role in soil formation processes through physical, chemical and biological processes. These affect stability of the volume of soil immediately surrounding the root as well as the geometry of soil pore spaces. Soil bacteria (e.g. *Rhizobium*) and especially fungi play important roles in soil structure of soil and soil aggregate stability, an effect enhanced by root exudates from healthy plants. Persistent, favourable soil structure is an important factor in improved plant health and agricultural production. There is evidence that soil-plant-microbe systems operate as a self-organising unit with soil microporosity serving as the driving force (40).

Aggregation determines pore structure and dispersion resistance of soils, and is a fundamental driver of soil and ecosystem functioning. Soil structure has many effects on soil fertility in terms of such physical processes concerning erosion, water evaporation, infiltration and runoff, internal drainage and aeration, and thermal properties. Soil structure also affects nutrient cycling and mineralisation, immobilisation, and ion exchange, as well as carbon cycling issues such as decomposition, humification, root and microbial biomass turnover, and most aspects of soil biological activities. The effect of soil structure on plant growth is therefore extremely important (14).

Arbuscular mycorrhizas affect soil aggregation through a range of mechanisms operating at the hierarchical levels of the plant community, the individual root, and the soil mycelium (68). These interactions affect soil water-holding capacity, infiltration and erosion; temporal and spatial distribution of anaerobic conditions; growth of plants and fungal hyphae; and the cycling of nutrients and carbon. Feedbacks between plants, soil microbes and soil structure regulate productivity and biogeochemical functioning of agroecosystems (23).

Microbes play well known roles in the breakdown of organic matter both on the soil surface as well as within the soil profile. Both these sources represent radiant energy "fixed" by the plant through photosynthesis into chemical energy in a wide range of substrates used to produce plant biomass. The return to the soil of this biomass as plant litter and/or crop residues provides the soil microbes with a range of organic compounds which are broken down to release energy for their use, including the recycling of key nutrients. As previously stated for every gram atomic weight of carbon taken up, 479 kJ of potential energy is obtained, with phytomass containing between 15 and 35 kJ /gram organic dry matter (most commonly 19-20 kJ g^{-1}: 44; 54).

Effects of addition of organic matter to soils and plants are highly significant (e.g. 14; 65), and interactions between plants and microbes result in such environmental effects as reduced requirement for fertilizers, irrigation water, herbicides and pesticides; greater groundwater recharge; reduced land degradation; increased sequestration of CO_2; and greater plant productivity (14).

Ultimately, roots and associated microbes can alter rhizosphere pH via redox related reactions. Soil pH is a master variable affecting the bioavailability of many nutrients and toxic elements and affects the physiology of roots and rhizosphere organisms. (41).

Aromatic acids such as salicylic and phthalic acids produced by microbes are absorbed onto soil exchange sites and alter physico-chemical character of soils. One such change is the decrease of *zeta* potential: the electrical potential developed at the soil solid-liquid interface. Gallic acid increases solubility of both calcium and manganese in soils by the formation of metal nutrient /gallic acid complexes and through redox reactions (86).

Cyclic and aromatic organic acids also play a range of roles in soils, such as weathering of materials, allelopathic interactions and inhibition of microbial growth. Vanillic, adipic and especially benzoic acids inhibit plant growth, and ferulic acid released from plant roots and microbe-mediated decomposition of organic matter may be involved in allelopathic interactions (86).

Recently, an alternate functional view of soil was proposed. In this view, the traditional two-dimensional concept of soil as a system composed of adsorbent (mineral surface) and adsorbate (soil organic matter) has been extended to a three-dimensional view of soil consisting of a multitude of largely independent micro-reactors formed around microbial cells, cell colonies and fungal hyphae. The soil mineral particles serve as components ("building blocks") for the construction of small micro-structures (i.e. microaggregates) which are built around microbial cells or cell colonies. Microbiota are known to actively create complex organo-clays (50). This view puts microbes literally at the centre of soil structure.

5.2. Effects on plant root architecture

The acquisition by roots of water and nutrients is of primary importance in basic plant biology and agriculture. Water and nutrient availability limit plant growth in all but a very few natural systems. The extensive use of irrigation, and of such chemicals as high analysis fertilizers, pesticides, fungicides, herbicides (16) and veterinary products affects soil microbes (14; 15) and in turn, plant function.

Plants are not merely passive recipients of soil nutrients. They integrate information about their immediate surroundings into such aspects as the allocation of resources to root production and proliferation; formation of symbioses with mycorrhizal fungi or N-fixing bacteria; alteration in rates of exudation; interactions with free-living bacteria; or the production of secondary defensive compounds. Since root morphology (or phenotype P) is genetically programmed (G) and environmentally (E) determined (i.e. G x E =P), signal transduction pathways must exist that interpret environmental conditions and lead to activation of genes enabling initiation of microbial symbioses or to initiation of lateral roots at a particular time and place. Microbial symbionts must also communicate their presence to their plant hosts, and plants have to distinguish between symbiont ("friend") and pathogen ("foe"). This suggests that reciprocal exchange of microbial signals is commonplace (11).

Root architecture refers to the spatial configuration of the root system. In general, studies consider only entire root systems, or at least a large subset of the root system of an individual plant, and not such fine structural details as root hairs. Its importance in plant productivity comes from the fact that many soil resources are unevenly distributed, or are subject to localised depletion. Thus, spatial deployment of the root system will strongly determine the ability of a plant to exploit these resources (56).

Plants respond to nutrient deficiency (and to excess) by altering root morphology, recruiting the help of microbes and changing the chemical environment of the rhizosphere. Components of RE (root exudates) assist plants in acquiring nutrients through acidifying or changing redox conditions and through direct chelation of nutrients (61). When grown under P-deficient conditions, white lupin (*Lupinus albus*) plants form cluster (or proteiod) roots that produce organic acids (especially citric – and malic acid) and mucilaginous polymers that acidify soils and solubilize bound phosphate. However, in order to do this, the presence of soil microbes is required for cluster root formation (35).

There is evidence that plants can "sense" the presence of nutrients and respond by allocating more resources to root growth as well as by directing root growth towards or within zones of relative nutrient abundance (61).

In the presence of protozoa, which prey extensively on (competing) soil bacteria, plants develop extensive, highly branched root systems due to stimulation of lateral root development: the result of hormonal influences of beneficial rhizobacteria (freed from competition by the protozoa) on root growth rather than nutrient effects. The resulting increased root surface allows more nutrients to be absorbed, and increases root exudation rates, thereby further stimulating bacterial-protozoan interactions. Thus, as well as stimulating flow of nutrients, protozoa promote a mutualistic interaction between plant roots and rhizobacteria. The ecological implications include the possibility that microbial root interactions could be seen as a microbial contest to gain maximum carbon (in many forms) from plants. From a practical perspective, the regulation of root architecture is a key determinant of both nutrient- and water use efficiency (11).

Mycorrhizae also affect root system development and hence root architecture. They exert positive effects resulting in more numerous and branched roots through effects on the mitotic cycle in root apical cells and increased formation of primordia. Negative- or no effects occur where root meristematic activity becomes obstructed by AMF colonisation (7).

Root longevity is positively correlated with mycorrhizal colonisation. This occurs through several mechanisms, including enhanced tolerance of drying soils and enhanced defence against root pathogens. Mycorrhizas are often associated with improved acquisition by roots of nutrients, especially P, and these common fungal partners may help reduce the deleterious effects of pathogens and root herbivory independent of their role in plant nutrition. In the context of climate change, interactions with soil biota could cause root life span in ecosystems to exhibit unforseen responses to climate change not predicted by general relationships between root properties and models of root efficiency (27).

The simple concept of the microbial loop (previously discussed) has been questioned since increased plant biomass resulting from protozoan-microbe interaction (by predation) is not always accompanied by increased plant uptake of N (excreted by protozoa). Instead, what is repeatedly observed is a dramatic change in root architecture, characterised by strong stimulation of growth of lateral roots where protozoa are present. This effect is also seen where plant growth promoting rhizobacteria are present; and the production by microbes of IAA seems a common mechanism for bacterial manipulation of root architecture (78).

More importantly, effects on root architecture of rhizobacteria seem largely mediated by protozoan grazing. Plants develop more extensive and highly branched root systems where protozoa are present, compared to where these grazers are absent. This grazing was found to promote auxin-producing rhizobacteria, which stimulates growth of the root system, allows enhanced uptake of soil nutrients, and increases exudation rates, thereby further stimulating bacterial-protozoan interactions (12; 32). Such tripartite protozoa-bacteria-plant interactions demonstrate the "marvellous complexity" of these interdependent ecological relationships (78).

However, studies of interactions between roots, bacteria and protozoa are difficult to conduct, given the large number of participants, the small physical scale of the interactions (78) and the fact that most soil microbes are unculturable. Hence, conclusions drawn are mostly inferred from "plant-soil" feedback studies. These studies focused upon how root exudates passively lost by the root system affect the activity and function of the entire microbial community. The interpretation of experimental results is seldom straightforward: plant roots can alter the physicochemical soil properties (such as nutrient availability) as well as microbial community, so soils being compared might be very different. Changes in physicochemical soil properties influence, in turn, microbial community, resulting in an additional feedback loop making it difficult to disentangle the range of physical, chemical and biological influences at work (42).

6. Plant production, and stress mitigation by microbes

"Stress" is considered to be a significant deviation from the optimal conditions for life. Wherever they grow plants are subjected to a great variety of stresses tending to restrict their growth, development and even survival. Stress disrupts normal structures and the coordination of various processes at the molecular, cellular and whole-plant levels. The restabilization and reparative counter-measures, the achievement of readjusted and adapted states, and the maintenance of greater powers of resistance to further stress all require additional energy and metabolites (54). These resources cannot be allocated to other processes such as plant growth and production and a loss of potential occurs in plants growing in stressed environments.

The widespread adoption today of greatly simplified crop sequences and a range of management interventions (e.g. tillage, soil amendments, use of biocides etc) has powerful effects on restructuring of microbial soil habitats such as the rhizosphere, soil aggregates, particulate organic matter and soil biopores. This leads to shifts in species abundance and diversity of soil biota (23). The ability of a crop plant to resist or tolerate insect pests and plant pathogens depends on optimal physical, chemical and especially biological properties of soils, where soils with active soil biology generally show good fertility as well as complex food webs and abundant competing microbes that prevent infection. It is suggested that soil nutrient availability (affected by soil microbes) not only affects the amount of damage that plants suffer from herbivores, but also the ability of the plant to recover from herbivory (4). This has important implications to plant health and productivity, since not all intensive production practices benefit plants.

Fungal endophytes present in roots and shoots of some plants enable plants to cope with such abiotic stress conditions as high temperature, drought and salinity (42). Proposed mechanisms of plant growth stimulation by the fungus *Trichoderma* include control of sublethal pathogens; environmental buffering against adverse soil pH, drought, cold, heat and waterlogging; solubilization of sparingly soluble nutrients like P and Mo; production of PGR which e.g. stimulate root production; degradation of allelochemicals; and buffering against soil toxins such as Al and Mn (64).

Drought stress is a major constraint to agriculture in the semi-arid tropics, and in many cases, symbiosis of crop plant with AM fungi increases host plant growth rates during drought stress and plant resistance to drought. Several mechanisms for this have been proposed, including production of AM of PGR; increased gas exchange and photosynthetic rates of plants; direct transport of water from soil via fungal hyphae; enhanced water uptake through improved hydraulic conductivity and increased leaf conductance; assimilation of nitrate by fungal hyphae; enhanced activity of plant enzymes affecting oxidative stress; regulation of osmosis in plants; and changes in plant cell-wall elasticity (76).

Worldwide, about 950 million ha of land is estimated to be salt-affected, with 23% of arable land affected by salinity and a further 10% by sodicity. In Australia, salinity affects about 17 million ha and sodicity a further 340 million ha (90). Soil salinity disturbs the plant-microbe interaction, which is a critical ecological factor tin promotion of plant growth in degraded ecosystems. In saline soils, PGPR improve plant health and performance through a range of mechanisms, including IAA production by salt-tolerant strains of some *Pseudomonas* spp. which also reduce plant uptake of toxic ions. This may result in stimulation of root growth and the reversal to some extent of the growth-retarding effects of salinity on shoots and roots. Plant stress can be reduced by presence of bacteria producing 1-aminocyclopropane-carboxylate (ACC) which leads to reduced levels of stress-related ethylene such that it does not become limiting to root growth. Also, beneficial bacteria can "engineer" positive interactions in the rhizosphere and stimulate plant growth, nutrient uptake, and symbiotic performance of *Rhizobium* under saline conditions (3; 25; 26; 38; 71).

External inputs into agricultural production systems include synthetic inorganic fertilizers (e.g. urea, ammonium nitrate, sulphates and inorganic phosphates); organic fertilizers (e.g. mineral ores, manure, fish emulsion, compost); pesticides (e.g. herbicides, fungicides, nematicides, insecticides); soil conditioners (e.g. lime, gypsum); veterinary health products (antibiotics, growth stimulants, antiparasitic products); and soil fumigants (e.g. chloropicrin, methyl isocyanate, propargyl bromide). Although these are applied with the aim of maximising productivity and economic returns, many of these inputs adversely affect soil microbes (15) as well as imposing physiological or metabolic stress on plants (54; 16).

Over-application of chemical fertilizers is often a serious issue. However, use of PGPR can contribute to reducing nutrient build-up in soils, and have a place in nutrient management strategies aimed at reducing fertilizer application rates and concomitant nutrient runoff from agricultural sources (46).

One effect of modern farming practices is the accumulation in soil of substances that decrease plant growth and harm soil biota (52; 92). Repeated application of glyphosate, as is commonly recommended, lowers the degradative potential of the indigenous microbes, resulting in soil pollution by relatively recalcitrant and difficult-to-hydrolyse organophosphonates (52). Free glyphosate herbicide in the soil profile alters composition of microbial communities, causing an "explosion" of populations of phytopathogenic fungi *Fusarium* and *Phytophthora* whilst reducing populations of highly beneficial fluorescent pseudomonads, Mn-reducing bacteria, and IAA-producing rhizobacteria. These effects result in reduced plant shoot and root biomass (92).

The weak point of introducing bacterial cells alone is that they have to rely on themselves to secure a niche in which they can develop, compete with indigenous microflora, and acquire sufficient available nutrients. This problem can be resolved by using bacteria in association with plants where bacteria form plant root biofilms to better survive under stress. They also are able to access RE as additional energy sources which may greatly intensify microbial degradation of glyphosate. In turn, organisms such as PGPR exert a wide range of beneficial effects on plant growth (through N fixation, P solubilisation, PGR production, improvement of mineral- and water status). Inoculation of sunflower seedlings growing in glyphosate-contaminated soil with *Enterobacter cloacae* K7 increased root length of 29% and development of extensively developed lateral roots, absent in control plants (52).

Besides degradation of glyphosate for use as an energy source, PGPR in polluted environments may protect the plant root system against colonisation by phytopathogenic fungi (52).

Other common soil contaminants are heavy metals (i.e. specific mass >5.0 g cm^{-3}), many of which (e.g. Hg, Cr-VI, Cd, Pb, Zn, Cu, Ni, As, Co, Sn, Mn) exhibit long-lasting toxicity in a wide range of environmental niches. These are of great concern because, unlike many other pollutants, they cannot be degraded either chemically or biologically. Though many of these cations are essentially required as trace elements and co-factors to carry out normal plant and microbial metabolic processes, they also display toxicity at low concentrations (i.e. 0.001 to 10 mg/l) with Hg and Cd metal ions even at concentrations of 0.001 to 0.1 mg/l. These toxicities strongly affect growth and metabolism of plants. Cadmium, for example, inhibits enzymic activities, DNA-mediated transformations, symbiosis between plants and microbes, and renders plants prone to fungal attack (2).

Bacteria including PGPR have evolved to resist heavy metal toxicity through multiple mechanisms resulting in physical sequestration, exclusion, complexation and/or detoxification of these elements. For example, bacterial extracellular polymers (e.g. proteins, polysaccharides, humic substances) effectively bind heavy metals and form a protective barrier surrounding the cell through a process called biosorption. Biofilms of such PGPR can effectively "cloak" the roots of plants as well. Moreover, siderophores secreted by many of the PGPR greatly reduce the bioavailability of the heavy metal, and hence its toxicity, by binding those metal ions with chemistry similar to that of Fe. Since these same PGPR also directly promote plant growth, their beneficial effects on plants growing in these soils demonstrate the potential of inoculation of PGPR to increase plant biomass. *Lupinus luteus* plants grown in fields with high levels of Cd, Pb and Cu were inoculated with *Bradyrhizobium* sp.750, *Pseudomonas* sp. and *Ochrobactrum cytisi*, resulting in increased plant biomass and tissue N content. In soils where the heavy metal content exceeds the limits of plant tolerance, it is often feasible to treat plants with PGPR, thereby stabilising and remediating ("bioremediating") metal-polluted soils , and allowing their revegetation (2; 71).

Fungi also act as biosorbents for metal accumulation, with hypomycetes and yeasts being highly effective sorbents for such cations as Fe, Z, Ni, Cu, Mo, Cr, Ag, and La. One kilogram of fungal biomass can absorb 63 mg of Cu; and a single gram of fungal mycelium as much as one gram of Zn (80). Humic compounds produced by soil microbial reaction with organic matter are also able to sequester a range of toxic and other elements, including heavy metals and radionuclides (44).

Soil microbes offer plant technologists and producers important potential benefits in situations where plants are growing under stressful conditions. These stresses decrease yield of crops and also restrict where crops can be grown. PGPR producing ACC-deaminase offer promise as bacterial inoculum for improvement of plant growth under such unfavourable environmental conditions as flooding, heavy metals, drought and salinity stress (71).

Interactions between AM fungi and bacteria are essential to plant production systems relying on biological processes, rather than agrochemicals, to maintain or improve soil fertility and plant health. The potential of microbial synergisms suggests excellent prospects for establishing multi-partite symbioses improving many agronomic traits. With evidence of physiological and genetic adaptation of microbes to their environment, locally sourced isolates are recommended for use in these biotechnological applications (76).

Plant-mediated decomposition and nutrient mineralisation via the rhizosphere is generally not seen as important to "conventional" agriculture: hence deliberate management of this process is seldom attempted. Clearly, however, greater reliance on plant-mediated mineralisation would reduce potential for nutrient losses through tight coupling between release (perhaps on plant demand) of soluble, potentially mobile nutrient forms and plant uptake of nutrients from the rhizosphere (23).

Many current crop management systems have created dependence on the many agricultural inputs centred on the root zone. This is similar to the pesticide treadmill of the 1960s of increased dependence on insecticides created by chemical control of insect pests. It is in the context of a highly modified soil environment and the cascading effects on soil biota that we view the agricultural rhizosphere and consider how best to redirect management in order to restore rhizosphere processes and agroecosystem function (23).

Rhizosphere regulation is regarded as an important agro-technology with great potential to reduce the adverse effects of agricultural production on the global environment and to manage highly stressed soils to allow the improvement of low productive lands to highly productive ones (35). Despite its great importance, we know remarkably little about rhizosphere functioning, and how best to manipulate it to our advantage (40).

Plants, assisted by the myriad of associated microbes with which they have coevolved, perform and provide many essential ecological services on which we all depend. These include the production of oxygen and regulation of atmosphere composition; fixation of solar energy and conversion into raw materials; purification of air and water; storage, cycling and global distribution of fresh water; decomposition of organic wastes and detoxification or sequestration of industrial pollutants; storage and cycling of nutrients; natural pest and disease control, provision of a genetic base (asset register) for food, fibre, pharmaceuticals and materials; and regulation of local and global climate.

Production systems based on microbial processes occurring in both rhizosphere and phyllosphere provide the basis for durable economies and the practical means to mitigate the wide range of "uninsurable risks" associated with the production of plant products under climatic extremes, using diminishing natural resources, and with ever increasing demands for fertile soil, water, food and other resources.

Table 1. Summary of abundance of organisms in the phyllosphere and rhizosphere (* mass in kg live weight: dry mass is 20 to 25% of masses shown)

Location	Organism	Abundance (M = million)	Reference
Phyllosphere	Bacteria	10^6 (1M) to 10^8 (100M) cells per cm^2 leaf surface	55
	Bacteria	46,000 to 25 M per gram wheat seeds	51
	Fungi	4,000-7000 per gram wheat seeds	51
	"Epiphytes"	1.5 M to 99M per gram green tissue	51
Rhizosphere	Bacteria	Total biomass of 5,000 to 7,000 kg/ha surface soil	51
	Fungi and algae	Total biomass of 2000-3000kg /ha surface soil	51
	Bacteria	10M to 10^{12} per gram of soil	40
	Bacteria	427 M per g soil (vs. 8.2 M in bulk soil) under beet	51
	Bacteria	11.3M per g (vs. 6.6 M in bulk soil) in clover	51
	Bacteria	653M per g (vs. 22.8M in bulk soil) in wheat	51
	Bacteria +actinomyctetes	10M to 100M per gram of soil	60
	Bacteria	Total biomass 1,100 kg/ha in top 15 cm of soil	60
	Bacteria	>100M culturable cells per gram of soil	63
	Bacteria	Total biomass* 400 to 5000 kg/ha in upper 15 cm soil	14
	Actinomycetes	Total biomass 1,100 kg/ha in upper 15 cm soil	60
	Actinomycetes	Total biomass* 400 to 5000 kg/ha in	14

		upper 15cm soil	
	Fungi	1,000 -100,000 propagules; 1,6 km hyphal length /g soil	60
	Fungi	5 to 25 km hyphal length per g soil	40
	Fungi	Total biomass 2,200 kg/ha in upper 15 cm soil	60
	Fungi	Total biomass 1,000 -15,000 kg/ha in upper 15 cm soil	14
	Algae	Total biomass* 10 to 500 kg/ha in upper 15 cm soil	14
	Protozoa	1,000 per g of soil	40
	Protozoa	Total biomass 20 to 200 kg/ha in upper 15 cm soil	14
	Nematodes	1,000 per g of soil	40
	Nematodes	Total biomass* of 10 to 150kg/ha in upper 15 cm soil	14

Literature cited.

1.Agrios, G. 2005. Plant pathology, 5th ed. Elsevier Academic Press, UK.

2. Ahemad, M. 2014. Remediation of metalliferous soils through the heavy metal resistant plant growth promoting bacteria: paradigms and prospects. Arab. J. Chem. http://dx.doi.org/10.1016/j.arabjc.2014.11.020 .Accessed online 25 May, 2015.

3. Ahemad, M. & M. Kibret. 2013. Mechanisms and applications of plant growth promoting rhizobacteria: current perspective. J. King Saud Univ. - Science (2014) 26:1-20. http://dx.doi.org/10.1016/j.jksus.2013.05.01 .Accessed online 25 May, 2015

4. Altieri, M. & C. Nicholls. 2003. Soil fertility management and insect pests: harmonizing soil and plant health in agroecosytems. Soil Till. Res. 72: 203-211.

5. Anderson, J. & K. Domsch. 1980. Quantities of plant nutrients in the microbial biomass of selected soils. Soil Sci. 130:211-216.

6. Atkinson, D. 2006. Arbuscular mycorrhizal fungi and the form and functioning of the root system. Pages 199-222 in: K. Mukerji, C. Manoharachary & J. Singh (eds.). 2006. Soil Biology: Volume 7. Microbial activity in the rhizosphere. Springer-Verlag. Berlin.

7. Artursson, V., R. Finlay & J. Jansson. 2005. Interactions between arbuscular mycorrhizal fungi and bacteria and their potential for stimulating plant growth. Environ. Microbiol. 8(1): 1-10. Doi: 10.1111/j.1462-2920.2005.00942.x.

8. Ball, A. 2006. Energy inputs in soil systems. Pages 79-89 in: N. Uphoff et al. (eds.). Biological approaches to sustainable soil systems. CRC Taylor & Francis Group. USA.

9. Balmer, D. & B. Mauch-Mani.2013. Small yet mighty- MicroRNAs in plant-microbe interactions. MicroRNA 2013. 2(1): 8 pp.

10. Boddey, R., B. Alves, V. Reis & S. Urquiaga. 2006. Biological nitrogen fixation in agroecosystems and in plant roots. Pages 177-189 in: N. Uphoff et al. (eds.). Biological approaches to sustainable soil systems. CRC Taylor & Francis Group. USA.

11. Bonkowski, M. 2004. Protozoa and plant growth: the microbial loop in soil revisited. New Phytol. 162: 617-631.

12. Bonkowski, M. & M. Clarholm. 2012. Stimulation of plant growth through interactions of bacteria and protozoa: testing the auxiliary microbial loop hypothesis. Acta Protozool. 51:237-247. Doi:10.4467/16890027AP.12.019.0765

13. Bonkowski, M., W. Cheng, B. Griffiths, J. Alphei & S. Scheu. 2000. Microbial-faunal interactions in the rhizosphere and effects on plant growth. Eur. J. Soil Biol. 36: 135-147.

14. Brady, N. & R. Weil. 2002. Soil Organic Matter. Pages 498-542 in: N. Brady & R Weil. The Nature and Properties of Soils 13th Ed. Prentice Hall. USA.

15. Bünemann, E., G. Schwenke & L.Van Zeiten. 2006. Impact of agricultural inputs on soil organisms- a review. Austral. J. Soil Res. 44: 379-406.

16. Chaboussou, F. 2004. Healthy crops: a new agricultural revolution. Jon Carpenter Publishing, UK.

17. Chakraborty, U., B. Chakraborty & A. Chakraborty. 2012. Induction of plant growth promotion in *Camellia sinensis* by *Bacillus megaterium* and its bioformulations. World J. Agric. Sci. 8(1): 104-112.

18. Cheng, W. & A. Gershenson. 2007. Carbon fluxes in the rhizosphere. Pages 31-56 in: Z. Cardon & J. Whitbeck (eds.). 2007. The rhizosphere: an ecological perspective. Elsevier Inc. USA.

19. Dakora, F. & D. Phillips. 2002. Root exudates as mediators of mineral acquisition in low-nutrient environments. Plant and Soil 245:35-47.

20. Dazzo, F. & Y. Yanni. 2006. The natural Rhizobium-cereal crop association as an example of plant-bacteria interaction. Pages 109-127 in: N. Uphoff et al. (eds.). Biological approaches to sustainable soil systems. CRC Taylor & Francis Group. USA.

21. De Meyer, G. & M. Höfte. 1977. Salicylic acid produced by the rhizobium *Pseudomonas aeruginosa* 7NSK2 induces resistance to leaf infection by *Botrytis cinerea* on bean. Phytopathology 87: 588-593.

22. Döring, T., M. Pautasso, R. Finckh & M.Wolfe. 2012. Concepts of plant health-reviewing and challenging the foundations of plant protection. Plant Pathol.61: 1-15.

23. Drinkwater, L. & S. Snapp. 2007. Understanding and managing the rhizosphere in agroecosystems. Pages 127-153 in: Z. Cardon & J.Whitbeck (eds.). The rhizosphere: an ecological perspective. Elsevier Inc. USA.

24. Duponnois, R. 2006. Bacteria helping mycorrhiza development. Pages 297-310 in: K. Mukerji, K., C. Manoharachary & J. Singh (eds.). Soil Biology: Volume 7. Microbial activity in the rhizosphere. Springer-Verlag. Berlin.

25. Egamberdieva, D. 2008. Plant growth promoting properties of rhizobacteria isolated from wheat and pea in loamy sand soil. Turk. J. Biol. 32: 9-15.

26. Egamberdieva, D. 2012. The management of soil quality and plant productivity in stressed environment with Rhizobacteria. Pages 27-40 in: D. Maheshwari (ed.). Bacteria in agrobiology: stress management. Springer-Verlag. Berlin.

27. Eissenstat, D., C. Wells, R. Yanai & K. Whitbeck. 2000. Building roots in a changing environment: implications for root longevity. New Phytol. 147: 33-42.

28. Ertürk, Y., S. Ercisli, A. Haznedar & R. Cakmakci. 2010. Effects of plant growth promoting rhizobacteria (PGPR) on rooting and root growth of kiwifruit (*Actinidia deliciosa*) stem cuttings. Biol. Res. 143: 91-98.

29. Gao, F., C. Dai & X. Liu. 2010. Mechanisms of fungal endophytes in plant protection against pathogens. Afr. J. Microbiol. Res. 4(13): 1346-1351.

30. Gardener, B. 2004. The ecology of *Bacillus* and *Paenibacillus* spp. in agricultural sustainability. Phytopathology 94: 1252-1258.

31. Gopalkrishanan, S., V. Srinivas, M. Vidya & A. Rathore. 2013. Plant growth-promoting activities of *Streptomyces* spp. in sorghum and rice. SpringerPlus 2013 (2): 574-583.

32. Griffiths, B., S. Christensen & M. Bonkowski. 2007. Microfaunal interactions in the rhizosphere: how nematodes and protozoa link above-and belowground processes. Pages 57-71 in: Z. Cardon & J.Whitbeck (eds.).The rhizosphere: an ecological perspective. Elsevier Inc. USA.

33. Habte, M. 2006. The roles of arbuscular mycorrizas in plant and soil health. Pages 129-147 in: N. Uphoff et al. (eds.). Biological approaches to sustainable soil systems. CRC Taylor & Francis Group. FL.

34. Hallmann, J., A. Quadt-Hallmann, W. Mahaffee & J. Kloepper. 1977. Bacterial endophytes in agricultural crops. Can. J. Microbiol. 43: 895-914.

35. Hashidoko, Y. 2005. Ecochemical studies of interrelationships between epiphytic bacteria and host plants via secondary metabolites. Biosci. Biotechnol. Biochem. 69(8): 1427-1441.

36. Hawkes, C., K. De Angelis & M. Firestone. 2007. Root interactions with soil microbial communities and processes. Pages 1-29 in: Z. Cardon & J.Whitbeck (eds.). The rhizosphere: an ecological perspective. Elsevier Inc. USA.

37. Hayat, S. & A. Ahmad (eds.). 2007. Salicylic acid: a plant hormone. Springer. The Netherlands.

38. Hayat, R., I. Ahmed & R. Sheirdil. 2012. An overview of plant growth promoting rhizobacteria (PGPR) for sustainable agriculture. Pages 557- 574 in: M. Ashraf et al. (eds.). Crop production for agricultural improvement. Springer Science + Business Media B.V.

DOI 10.1007/1007/978-94-007-4116-4_22.

39. Hidayat, C., D. Arief, A. Nurbaity & J. Sauman. 2013. Rhizobacteria selection to enhance spore germination and hyphal length of arbuscular mycorrhizal fungi *in vitro.* Asian J. Agric. Rural Dev. 3(4):199-204.

40. Hinsinger, P., A. Bengough, D. Vetterlein & I. Young. 2009. Rhizosphere: biophysics, biogeochemistry and ecological relevance. Plant Soil 321: 117-152.

41. Hinsinger, P., C. Plassard, C. Tang & B. Jaillard. 2003. Origins of root-mediated pH changes in the rhizosphere and their response to environmental constraints: a review. Plant Soil 248: 43-59.

42. Hodge, A. & A. Fitter. 2013. Mechanisms of plant competition. Microbial mediation of plant competition and community. Funct. Ecol. 27: 865-875. doi: 10.1111/1365-2435.12002.

43. Hoyle, F. & D. Murphy. 2006. Seasonal changes in microbial function and diversity associated with stubble retention versus burning. Austral. J. Soil Res. 44: 407-423.

44. Jackson, W. 1993. Humic, fulvic and microbial balance: organic soil conditioning. Jackson Research Center. USA.

45. Johnson, N. & C. Gehring. 2007. Mycorrhizas: symbiotic mediators of rhizosphere and ecosystem processes. Pages 73-100 in: Z. Cardon & J.Whitbeck (eds.). The rhizosphere: an ecological perspective. Elsevier Inc. USA.

46. Kaymak, H. 2010. Potential of PGPR in agricultural innovations. Pages 45-80 in: D. Maheshwari (ed.). 2010. Plant growth and health-promoting bacteria. Microbiology monograph 18. Springer-Verlag, Berlin.

47. Keen, B. & T. Vancov. 2010. *Phytophthora cinnamomi* suppressive soils. Pages 239-250 in: A. Mendez-Vilas (ed.). Current research, technology and education topics in applied microbiology and microbial biotechnology. Formatex microbiology series No. 2, vol. 1. Formatex, Spain.

48. Khalid, A., M. Arshad, & Z. Zahir. 2006. 2006. Phytohormones: microbial production and applications. Pages 207-220 in: N. Uphoff et al. (eds.). Biological approaches to sustainable soil systems. CRC Taylor & Francis Group. USA. .

49. Kishore, G., S. Pande & A. Podile. 2005. Phylloplane bacteria increase seedling emergence, growth and yield of field-grown groundnut (*Arachis hypogaea* L.). Lett. Appl. Microbiol. 40(4): 260-268.

50. Kleber, M. 2010. Minerals and carbon stabilisation: towards a new perspective of mineral-organic interactions in soils. Pages 77-79 in: Proceedings of the 19th World Congress of Soil Science, Soil solutions for a changing world. 1-6 August, Brisbane, Australia.

51. Krasil'nikov, N. 1961. Soil microorganisms and higher plants. The National Science Foundation, Washington DC (translated from Russian by Dr. Y. Halperin, Israel Program Scientific trans. Jerusalem).

52. Kryuchkova, Y., G. Burygin, N. Gogoleva, Y. Gogolev, M. Chernyshova, O. Makarov, E. Federov & O. Turkovskaya. 2013. Isolation and characterisation of a glyphosate-degrading rhizosphere strain, *Enterobacter cloacae* K7. Microbiol. Res. 169: 99-105.

53. Kumar, A. & H. Bais. 2012. Wired to the roots. Impact of root-beneficial microbe interactions on aboveground plant physiology and protection. Plant Sig. Behav. 7(12): 1598-1604.

54. Larcher, W. 2003. Physiological plant ecology: ecophysiology and stress physiology of functional groups.4th ed. Springer-Verlag. Berlin.

55. Lindow, S. & M. Brandl. 2003. Microbiology of the phyllosphere. Appl. Environ. Microbiol. 69(4): 1873-1883. doi:10.1128/AEM.69.4.1875-1883.2003.

56. Lynch, J. 1995. Root architecture and plant productivity. Plant Physiol. 109: 7-13.

57. Manoharachary, C. & K. Mukerji. 2006. Rhizosphere biology- an overview. Pages 1-15 in: K. Mukerji, C. Manoharachary & J. Singh (eds.). Soil Biology: Volume 7. Microbial activity in the rhizosphere. Springer-Verlag. Berlin.

58. Madhaiyan, M., S. Poonguzhali, M. Senthilkumar, S. Seshadri, H. Chung, J. Yang, S. Sundaram & T.Sa. 2004. Growth promotion and induction of systemic resistance in rice cultivar Co-47 (*Oryza sativa* L.) by *Methylobacterium* spp. Bot. Bull. Acad. Sin. 45:315-324.

59. Marschner, P. & S. Timonen. 2006. Bacterial community composition and activity in rhizosphere of roots colonised by arbuscular mycorrhizal fungi. Pages 139-154 in: K. Mukerji, C. Manoharachary & J. Singh (eds.). Soil Biology: Volume 7. Microbial activity in the rhizosphere. Springer-Verlag. Berlin.

60. Maser, C., A. Claridge & J. Trappe. 2010. Trees, truffles, and beasts: how forests function. Rutgers University Press, USA.

61. McNear, D. 2013. The rhizosphere-roots, soil and everything in between. Nat. Edu. Knowl. 4(3): 1-13.

62. Mechaber, W., D. Marshall, R. Mechaber, R. Jobe & F. Chew. 1996. Mapping leaf surface landscapes. Proc. Natl. Acad. Sci. USA 93: 4600-4603.

63. Nehl, D., G. Oliver & G. Knox. 2006. Significance of bacteria in the rhizosphere. Pages 89-119 in: K. Mukerji, C. Manoharachary & J. Singh (eds.). Soil Biology: Volume 7. Microbial activity in the rhizosphere. Springer-Verlag. Berlin.

64. Neumann, B. & M. Laing. 2006. *Trichoderma*: an ally in the quest for soil system sustainability. Pages 491-500 in: N. Uphoff et al. (eds.). Biological approaches to sustainable soil systems. CRC Taylor & Francis Group. USA.

65. Phelan, P. 2004. Connecting belowground and aboveground food webs: the role of organic matter in biological buffering. Pages 250-284 in: F. Magdoff & R. Weil (eds.). Soil organic matter in sustainable agriculture. CRC Press LLC. USA.

66. Philips, D., H. Ferris, D. Cook & D. Strong, 2003. Molecular control points in rhizosphere food webs. Ecology 84: 816-826.

67. Ramey, B., M. Koutsoundis, S. von Bodman & C. Fuqua. 2004. Biofilm formation in plant-microbe associations. Curr. Opin. Microbiol. 7: 602-609.

68. Rillig, M. & D. Mummey. 2006. Mycorrhizas and soil structure. New Phytol. 171:41-53.

69. Römheld, V. & G. Neumann. 2006. The rhizosphere: Contributions of the soil-root interface to sustainable soil systems. Pages 91-108 in: N. Uphoff et al. (eds.). Biological approaches to sustainable soil systems. CRC Taylor & Francis Group. USA.

70. Rovira, A. 1969. Plant root exudates. Bot. Rev. 35(1): 35-57.

71. Saharan, B. & V. Nehra. 2011. Plant growth promoting rhizobacteria: a critical review. Life Sci. Med. Res. 21. 30 pp. http://astonjournals.com/lsmr . Accessed online 25 May, 2015.

72. Saxena, A., R. Shende & M. Grover. 2006. Interactions among beneficial organisms. Pages 121-137 in: K. Mukerji, C. Manoharachary & J. Singh (eds.). Soil Biology: Volume 7. Microbial activity in the rhizosphere. Springer-Verlag. Berlin.

73. Sbrana, C. 2006. Fungal recognition responses to host-derived signals by arbuscular mycorrhizal fungi. Pages 223-243 in: K. Mukerji, C. Manoharachary & J. Singh (eds.). Soil Biology: Volume 7. Microbial activity in the rhizosphere. Springer-Verlag. Berlin.

74. Seneviratne, G., M. Weerasekara, K. Seneviratne, J. Zavahir, M. Kecskés & I. Kennedy. 2010. Importance of biofilm formation in plant growth promoting rhizobacterial action. Pages 81-97 in: D. Maheshwari (ed.). 2010. Plant growth and health-promoting bacteria. Microbiology monograph 18. Springer-Verlag. Berlin.

75. Shahab, S., N. Ahmed & N. Khan. 2009. Indole acetic acid production and enhanced plant growth promotion by indigenous PSBs (phosphate solubilizing bacteria). Afr. J. Ag. Res. 4(11): 1312-1316.

76. Shtark, O., A. Borisov, V. Zhukov, N. Provorov & I. Tikhonovich. 2010. Intimate associations of beneficial soil microbes with host plants. Pages 119-196 in: G. Dickson & E. Tilston (eds.). Soil microbiology and sustainable crop production. Springer Science + Business Media B.V.

77. Shweta, B., D. Maheshwari, R. Dubey, D. Arora, V. Bajpai & S. Kang. 2008. Beneficial effects of fluorescent pseudomonads and seed germination, growth promotion, and suppression of charcoal rot in groundnut (*Arachis hypogea* L.). J. Microbiol. Biotechnol. 18(9): 1578-1583.

78. Spaepen, S., J. Vanderleyden & R. Remans. 2007. Indole-3-acetic acid in microbial and microorganism-plant signalling. FEMS Microbiol. Rev. (2007):1-24. doi:10.1111/j.1574-6976.2007.00072.x.

79. Stefan, M., N. Munteau, V. Stoleru & M. Mihasan. 2013. Effects of the inoculation with plant growth-promoting rhizobacteria on photosynthesis, antioxidant status and yield of runner bean. Rom. Biotech. Lett. 18(2): 8132-8143.

80. Sterflinger, K. 2000. Fungi as geologic agents. Geomicrobiol. J. 17: 97-104.

81. Thies, J. & J. Grossman. 2006. 5. The soil habitat and soil ecology. Pages 59-78 in: N. Uphoff et al. (eds.). Biological approaches to sustainable soil systems. CRC Taylor & Francis Group. USA.

82. Trappe, J. & A. Claridge. 2010. The hidden life of truffles. Scientific Amer. April: 78-84.

83. Toro, M., R. AzcÓn & J. Barea. 1997. Improvement of arbuscular mycorrhiza development by inoculation of soil with phosphate-solubilizing rhizobacteria to improve rock phosphate bioavailability (^{32} P) and nutrient cycling. Appl. Environ. Microbiol. 63(11): 4408-4412.

84. Turner, B., E. Frossard & A. Oberson. 2006. Enhancing phosphorus availability in low-fertility soils. Pages 191-205 in: N. Uphoff et al. (eds.). Biological approaches to sustainable soil systems. CRC Taylor & Francis Group. USA.

85. Vastakaite, V. & O. Buzaite. 2001. Plant growth promoting properties of culturable phylloepiphytic pine bacteria. Biologija 57(1): 33-36.

86. Vranova, V., K. Rejsek & P. Formanek. 2013. Aliphatic, cyclic, and aromatic organic acids, vitamins, and carbohydrates in soil: a review. Sci. World. J. Volume 2013, Article ID 524239. 15 pp. http://dx.doi.org/10.1155/2013/524239. Accessed online 25th May, 2015.

87. Walker, T., H. Bais, E. Grotewald & J. Vivanco. 2003. Root exudation and rhizosphere biology. Plant Physiol. 132: 44-51. www.plantphysiol.org/Cgi/doi/10.1104/pp.102.019661 Accessed online 25th May, 2015.

88. Weiser, C., H. Quamme, E. Proebsting, M. Burke & G. Yelenosky. 1979. Plant freezing injury and resistance. Pages 55-84 in: B. Barfield & J. Gerber (eds.). Modification of the aerial environment of plants. ASAE Monograph No.2. American Society of Agricultural Engineers. USA.

89. Whipps, J., P. Hand, D. Pink & G. Bending. 2008. Phyllosphere microbiology with special reference to diversity and plant genotype. J. Appl. Microbiol. 105: 1744-1755.

90. Wong, V., R. Greene, R. Dalal & D. Murphy. 2010. Soil carbon dynamics in saline and sodic soils: a review. Soil Use & Manag. 26: 2-11.

91. Yeates, G. & T. Pattison. 2006. Moving up within the food web: protozoa and nematodes. Pages 149-162 in: N. Uphoff et al. (eds.). Biological approaches to sustainable soil systems. CRC Taylor & Francis Group. USA.

92. Zobiole, L., R. Kremer, R. Oliviera & J. Constantin. 2010. Glyphosate affects micro-organisms in rhizospheres of glyphosate-resistant soybeans. J. Appl. Microbiol. 100: 118-127.

SEED TREATMENTS WITH VOLATILE OILS INDUCE DROUGHT RESISTANCE IN GRAIN AND VEGETABLE SEEDLINGS

Joshua D. Klein, ARO-Volcani Center, Ministry of Agriculture, Bet Dagan, Israel

Amanda Firmansayah, Nurhaya Panga, Faculty of Agriculture, University of Hassanudin, Makassar, Indonesia

Mariana Peil da Rosa, Faculty of Agronomy, Universidade Federale de Pelotas, Pelotas, Argentina

Seed treatments are an established means of protecting seedings against biotic stress (fungal, bacterial, or insect attack) and more recently against abiotic stress such as chilling. Almost all of the compounds used as seed treatments are synthetic, and many of them are dangerous for humans to handle. We investigated the use of thymol and carvacrol, two essential oils extracted from Origanum syriacum (wild oregano), as treatments for carrot, radish, wheat, and barley seeds. The oils were applied to seeds both in solution and by means of solid matrix priming, and were compared with trinexapac-ethyl, a synthetic gibberellin-synthesis inhibitor known to induce drought resistance. Solid matrix priming did not confer any benefits, and in fact often promoted fungal and bacterial infections in the seeds. Soaking seeds in volatile oil solutions substantially reduced bacterial and fungal load on the seeds, enhanced water retention by droughted seedlings by increasing leaf relative water content, leaf thickness, and cell membrane stability, and induced pigmentation related to protective antioxidant activity. Germination and growth in some seedlings was decreased at higher concentrations of volatile oils or trinexapac-ethyl. Careful pre-testing of compounds at different concentrations with different seeds will allow farmers to benefit from natural growth regulators that not only provide protection against biotic and abiotic stresses to seedlings but also enhance the nutritional value of the plants for consumers.

HORMONAL APPROACHES IN INFLUENCING RUBBER PRODUCTION OF TARAXACUM KOK-SAGHYZ ROOTS

Muhammad Akbar Abdul Ghaffar[1] Tea Meulia[2] and Katrina Cornish[1]

[1]*Department of Horticulture and Crop Science, The Ohio State University, OARDC, Williams Hall, 1680 Madison Avenue, Wooster, OH 44691*

[2]*Molecular and Cellular Imaging Center, OSU/OARDC, Selby Hall, 1680 Madison Avenue, Wooster OH 44691*

Abstract

Natural rubber (NR) is one of the world's important commodities and research on improving latex production will benefit the NR industry from the rubber-producers (upstream) to the manufacturers and consumers (downstream). In Hevea rubber production, ethylene has been used to prolong latex flow for improved yield, and recently methyl jasmonate has shown similar results. The same methods may prove effective in other rubber-producing plants such as Taraxacum kok-saghyz (TK). The objective of this study is to evaluate the use of hormonal stimulants on TK latex production in order to gain a better understanding of rubber particle production. TK roots were submerged in ethephon and methyl jasmonate solutions with and without environmental stimuli, i.e. cold temperature induction. Rubber per plant was measured using Accelerated Solvent Extraction (ASE) and changes in the roots, laticifers and rubber particles were observed with near infrared (NIR) spectroscopy as well as with confocal, electron and light microscopy. The results show that the treatment of 1% ethephon solution at ambient temperature increased rubber concentration significantly ($P<0.05$) after nine days, with 81.9±4.3mg/g in comparison to the distilled water control level of 50.7±4.4mg/g, and the freshly harvested control (0 days) level of 54.0±2.8mg/g. Furthermore, the rubber particles observed under transmission electron microscope (TEM) show different morphologies in ethephon compared to the control. These results and observations indicate hormone applications may increase rubber concentration in TK and also affect the rubber particle ontogeny within the laticifer.

Introduction

The bark of *Hevea* trees contains pipe-like vessels called laticifers which produce latex (de Faÿ and Jacob, 1989). Laticifers are also found in some other rubber producing plants, such as *Ficus elastica* and *Taraxacum kok-saghyz* (TK) (Artschwager and McGuire, 1943; Metcalfe, 1967; Esau, 1977). Generally, latex from *Hevea* trees is collected from a scheduled tapping activity where a bark incision is made every two to three days using a special knife. During tapping, a unique tree stimulation technique is employed by appling ethephon paste or gaseous ethylene to the bark incision to obtain better rubber yields (Abraham *et al.*, 1968; Yew *et al.*, 1990). This practice prolongs latex flow and increases productivity with respect to the amount of latex collected or the dry rubber content. Recently, the application of jasmine oil, which contains methyl jasmonate, to the tapping panel of *Hevea* trees has been shown to increase rubber production as well. However, the downfall of the ethephon stimulatory practice is tapping panel dryness which can be induced by too high frequency or concentration of the ethephon application. The approach of using ethephon or ethylene on other rubber producing plants in order to increase rubber production is still novel, yet has vast potential to be developed as part of routine agronomic practices or post-harvest treatments.

Thus, our study was conducted to evaluate the use of these hormonal stimulants on rubber production in TK, an alternative rubber producing plant that is currently being developed as a commercial crop in the US and Europe (Buranov and Elmuradov, 2010; Cornish *et al.*, 2013). The study also includes cold temperature induction as an environmental stimulus as it was found in a previous study that cold storage increases rubber yield in TK roots (Cornish *et al.*, 2013). We hypothesize that the amount of TK rubber will increase with the use of hormones and/or with cold temperature treatment of the roots. Here, we focus on changes observed in the rubber concentration and rubber particle ontogeny in TK after the application of ethephon under two different temperatures. Previously, smaller scale experiments were conducted using ethephon and methyl jasmonate and our observations of rubber concentration, spectroscopy changes using near infra-red (NIR) spectroscopy and microscopy of the laticifers indicated that the ethephon treatment was superior to the methyl jasmonate treatment. The knowledge gained from the current study will assist in identification of the *de novo* site of rubber particle ontogeny in TK, while relating increases in rubber concentration to rubber particle ontogeny. Also, with more research, TK may become a useful model plant for understanding the occurrence of tapping panel dryness in *Hevea* trees.

Materials and Methods

One-year old TK plants were selected randomly from plants grown from seedlings transplanted into outdoor raised beds, at a density of 400,000 plants/acre, with 30 plants distributed per treatment. The plants were harvested, washed in water, and the roots then were soaked in four different treatments: distilled water (control) or 1% ethephon at either ambient temperature or 4°C. After 9 days, roots were harvested and resin and rubber content quantified from 0.25 grams dried and ground root bagasse by accelerated solvent extraction (ASE) (Pearson *et al.*, 2013). In order to account for increases in acetone soluble materials (resins) observed in the roots, samples were subjected to two solvent sequential extractions, where acetone was first used to remove the resins, and hexane was then used to extract the rubber. As a reference, 30 fresh plants (0 day) were harvested and the roots directly dried and their rubber quantified by ASE. In addition, ten plants per treatment were chosen at random for latex quantification (LQ) before rubber quantification as described (Cornish et al., 1999) with minor modification. Briefly, three reps of 1 ml TK root homogenate were analyzed for each sample. Each 1 ml sample was centrifuged for 5 minutes at 17,000g. The latex rose to the top of the centrifuge tubes and was coagulated using glacial acetic acid, re-centrifuged, collected, dried and then weighed. Three separate 1 ml homogenate samples were dried and weighed to measure non-latex solids within the root homogenates. Latex is the sub-fraction of rubber that is still in the form of individual particles in aqueous cytosol. Rubber is the sum of the latex fraction and the rubber fraction which has coagulated/solidified within the roots before extraction. For rubber particle ontological observation, the root tissue samples underwent primary fixation in 3% glutaraldehyde, 2% paraformaldehyde and 0.1 M phosphate buffer. Samples were then post-fixed in 1% osmium tetroxide and 1% uranil acetate, dehydrated by a series of ethanol applications, and later infiltrated with propylene and resin. The resulting tissue blocks were then embedded with EM Bed-812 resin and after incubating, the blocks were cut at 70nm thickness using a LEICA EM-UC6 Ultra Microtome. Lead citrate stain was used to stain the sectioned tissues for further observation under the microscope.

Results

a) *Rubber and latex*

After 9 days, the highest root rubber concentration (mg/g) was 81.9±4.3 mg/g in roots treated with 1% ethephon solution at ambient temperature in comparison to 50.7±4.4 mg/g in the control roots, which were treated with distilled water at ambient temperature (Fig 1), and the fresh roots (0 days) rubber concentration of 54.0±2.8mg/g. The second highest rubber-inducing treatment was 1% ethephon at 4°C, with 63.4±3.5 mg/g after 9 days. Cold treated roots without ethephon had the lowest rubber concentration at 47.3±3.1 mg/g, but this was not significantly different from the control plants. The latex quantification results indicate that 1% ethephon solution at ambient temperature yields the highest concentration of latex/dry weight root with 20.8±4.7 mg/g in comparison the control of 15.0±4.8 mg/g latex (Fig 2). The resin concentration obtained from samples in every treatment indicates that regardless of the treatment (distilled water or ethephon) ambient temperature results in higher resin concentration in comparison to cold treated plants. Average resin levels were about 30 mg/g for ambient temperature compared to only an average level of 19 mg/g in the cold treatment.

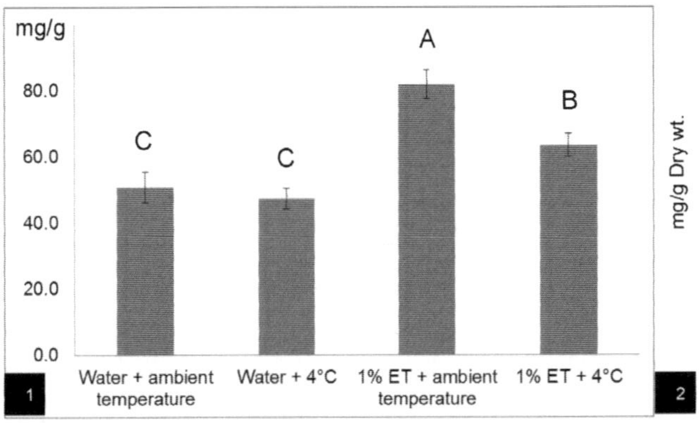

Figure 1: Total rubber concentration per root (mg/g) by ASE of four different treatments. Means with the same letter(s) are not significantly different at $p<0.05$ (LSD test)

Micro

b) Microscopy observation

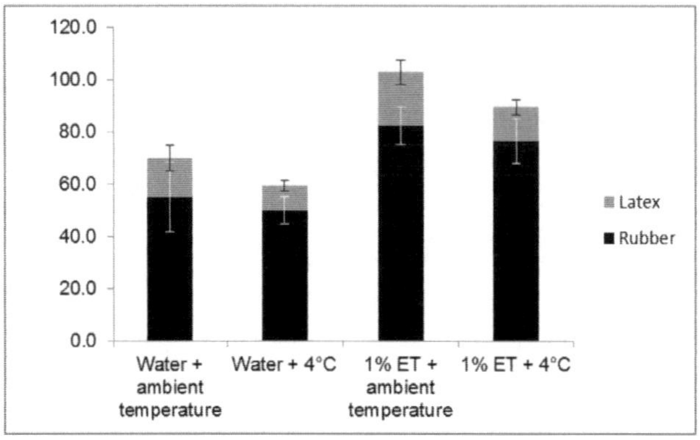

Figure 2: Root latex and rubber concentrations (mg/g) obtained from the LQ and then ASE of 10 plants samples from each treatment,

Most of the rubber particles observed from the four treatments are spherical or ovoid in shape similar to those observed in fresh TK roots (not shown). In fresh TK roots, it was observed that most of these rubber particles appeared in the cytoplasm. Interestingly, in this study, the cytoplasm was affected significantly by the different treatments; the cytosol became condensed and shrunk, probably due to aging in cold temperature and in both cold and ambient temperature of ethephon treatments. There are also other observable changes; the swelling of endoplasmic reticulum (ER) based on the many dense particles (Ishikawa, 1996) and numerous polyribosomes (Čiamporova and Mistrik, 1993), and the destroyed and digested cell organelles within the cytosol. In our study, the cytoplasm and the rubber particles in the control treatment (distilled water at ambient temperature, Fig 3a) appeared similar to those in the fresh TK roots (not shown) except that there were many vesicles apparent, especially around the vacuoles. However, in the cold treated plants, the cytoplasm shrank and the ER looks swollen with the establishment of denser particles (Fig 3b). Meanwhile, organelles clearly visible found in fresh TK plants cannot be distinguished in plants treated with ethephon at ambient temperature (Fig 3c). Within these cells, the cytosol appeared empty as only rubber particles with monolayers were clearly distinct. Furthermore, most of the cell walls from this treatment had deteriorated. In certain laticifer cells, groups of rubber particles were found within senescing cytoplasm (Fig 3c). In contrast with ambient temperature treatments (Fig 3c), a denser cytoplasm and ER could be detected clearly in

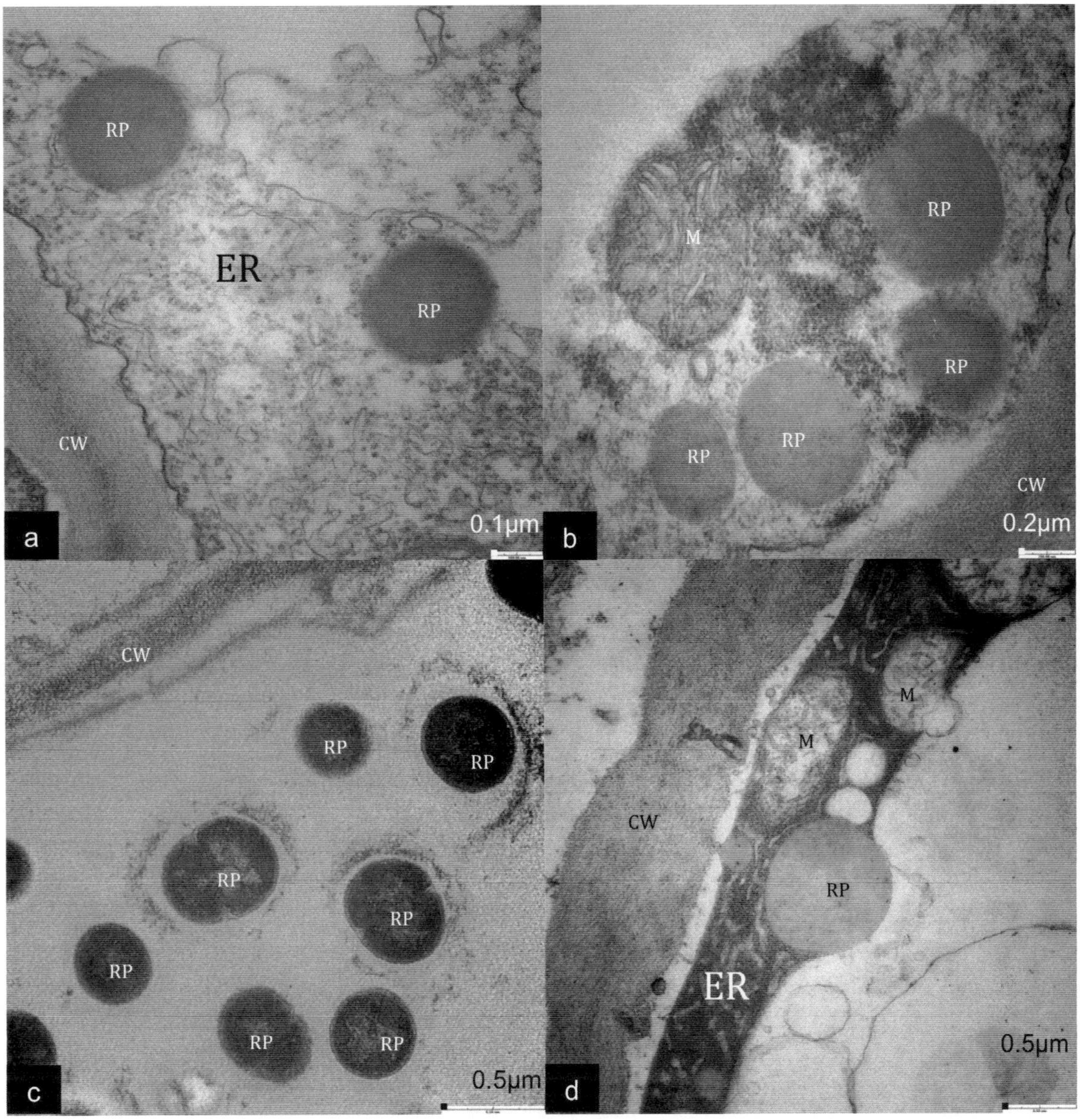

Fig 3. Rubber particles formed under different treatments **a** distilled water at ambient temperature **b** distilled water at 4°C **c** 1% ethephon at ambient temperature **d** 1% ethephon at 4°C
Cell wall (CW) Mitochondria (M) Rubber particles (RP) Endoplasmic reticulum (ER)

the ethephon at cold temperature treatment (Fig 3d). Further observation of this treatment indicates that the cytoplasm was shrinking inward away from the cell wall, and degrading as seen from damaged mitochondria as well as smaller vesicles appearing near the compacted ER.

Discussion

Ethephon caused an increase in rubber concentration in the treated roots, over 9 days, suggesting a direct effect on rubber biosynthesis. However, no clear results were obtained during this time with cold treatment. This is perhaps not surprising since rubber concentration did not increase until 45 days of cold treatment in fresh roots of ample size

(>7g) stored at 4°C post-harvest. Ethylene may have been involved in the observed response but atmospheric levels in storage were not measured. The delay in synthesis does suggest either that ethylene levels took several weeks to build to an inductive level or that the storage venting was sufficient to prevent ethylene build up and allowed a cold-induction alone. It has previously been shown that 35 to 45 days of 5°C nights are required to induce rubber biosynthesis in guayule plants (Cornish *et al.*, 2003) and the TK system may respond similarly. However, further studies on the ethylene released from the stored TK roots are needed before this can be confirmed.

Throughout this study, we have categorized the rubber particles based on their location in the cell. In the fresh root, rubber particles located in the vacuoles are usually spherical or ovoid with a smooth perimeter (not shown). Rubber particles located inside the cytoplasm, the cytoplasmic rubber (Backhaus and Walsh, 1983), generally have an irregular perimeter (not shown). The cytoplasmic rubber observed in fresh TK roots appears to be generated inside a laticifer plastid similar to those reported in *Papaver bracteatum* (Nessler and Mahlberg, 1978). We have not yet characterized the laticifer plastids from ethephon and cold but morphologically there are no irregular shaped rubber particles observed either in the fresh or the aging cytosol. We do find empty vesicles that we believe result from the rubber being pinched out from the cytoplasm; the same mechanism that was reported in *Parthenium argentatum* (Backhaus and Walsh, 1983). This may indicate that soaking the roots in the solutions affected rubber particle ontogeny.

During this experiment, both cytoplasm and ER underwent morphological changes which differed with the different treatments. All images of the ethephon and cold treated roots indicate that their cytoplasm shrank. The shrinking of cytoplasm is likely due to destructive processes that parallel digestion of other organelles in the cytosol (Ishikawa, 1996). The treatment resulting in roots with the highest rubber concentration (ethephon at ambient temperature) revealed that rubber particles were present without most of the typical organelles being observed nearby. Meanwhile, the second highest treatment (ethephon at 4°C) showed osmiophilically denser and darker cytoplasm. The mitochondria in this treatment degraded and the ER vesicles became obvious as the cytoplasm shrank and detached itself from the cell wall. We suggest that ethephon at ambient temperature stimulates the rate of cytoplasmic senescence and that the cytoplasm breakdown products are channeled into rubber particle biosynthesis. However, when cold temperature is coupled with ethephon, senescence is delayed, which explains why this treatment resulted in the second highest rubber levels. Meanwhile, in the treatment with the lowest rubber levels (distilled water at 4°C); we suggest that the cell senescing process is even slower. Without ethephon assisting the aging process in the cytosol, the cell is only experiencing cell injury due to cold conditions which does not lead to new rubber particle synthesis.

Conclusion

The ethephon treatment developed to enhance *Hevea* rubber production may have the potential to be expanded to other rubber producing plants. Also, cold treatment and the use of ethephon on TK plants affect rubber particle ontogeny, based on the dynamic changes materializing in the cytoplasm.

Acknowledgements We would like to thank Sarah K. McNulty (Department of Horticulture and Crop Science, OSU/OARDC) for proof reading the report and Karolyne Kaszas (Molecular and Cellular Imaging Center, OSU/OARDC) in helping us on preparing the samples for microscopic observation.

References

Abraham, P.D., Wycherley, P. R. and Pakianathan, S. W. (1968) Stimulation of Latex Flow in *Hevea brasiliensis* by 4-Amino-3, 5, 6-Trichloropicolinic acid and 2-Chloroethanephosphonic acid, J. Rubb. Res. Inst. Malaya 20(5), Rubber Research Institute of Malaysia, Kuala Lumpur: 291-305

Artschwager, E and McGuire, R.C. (1943) Contribution to the Morphology and Anatomy of the Russian Dandelion (*Taraxacum kok-saghyz*), Tech. Bull. (843):1-24

Addicott, F.T (1944) A differential stain for rubber in guayule, Stain Technol. 19:99-102

Backhaus, R.A. and Walsh, S. (1983) The Ontogeny of Rubber Formation in Guayule, *Parthenium argentatum* Gray Vol.144 (3):391-400

Buranov, A.U. and Elmuradov, B.J. (2010) Extraction and Characterization of Latex and Natural Rubber from Rubber-Bearing Plants, Journal of Agricultural and Food Chemistry Vol.58 (2): 734-743

Čiamporova, M. and Mistrik, I. (1993) The Ultrastructural Response of Root Cells to Stressful Conditions, Environmental and Experimental Botany, Vol. 33(1): 11-26

Cornish, K., Chapman, M.H., Nakayama, F.S., Vinyard, S.H. and Whitehand, L.C. (1999) Latex Quantification in Guayule Shrub and Homogenate, Industrial Crops and Products (10): 121-136

Cornish, K. and Backhaus, R.A. (2003) Induction of Rubber Transferase Activity in Guayule (*Parthenium argentatum* Gray) by Low Temperatures, Industrial Crops and Products 17: 83-92

Cornish, K., Bates, G.M., McNulty, S.K., Kopicky, S.E., Grewal, S., Rossington, J., Michel, F.C.Jr., Walker, S. and Kleinhenz, M.D. (2013) Buckeye Gold Storage: A Study into Rubber Production in *Taraxacum kok-saghyz* with an Emphasis on Post-harvest Storage, USA Tire Technology International (10): 36-38 de Faÿ, E. and Jacob, J.L. (1989) Anatomical Organization of the Laticiferous System in the Bark in Physiology of Rubber Tree Latex pp.4- 14

Esau, K. (1977) Secretory Structures in Anatomy of Seed Plants 2nd Ed. pp. 308-335

Ishikawa, H.A. (1996) Ultrastructural Features of Chilling Injury: Injured Cells and the Early Events during Chilling of Suspension-Cultured Mung Bean Cells, American Journal of Botany, Vol. 83, No. 7: 825-835

Metcalfe, C.R. (1967) Distribution of Latex in Plant Kingdom, Economic Botany Vol.21 (2): 115-127

Nessler, C.L. and Mahlberg, P.G. (1978) Laticifer Ultrastructure and Differentiation in Seedlings of *Papaver bracteatum* Lind., Population Arya II (Papaveraceae), American Journal of Botany,65:978-983

Pearson, C.H., Cornish, K. and Rath, D. J. (2013) Extraction of Natural Rubber and Resin from Guayule using an Accelerated Solvent Extractor, Industrial Crops and Products Vol. 43:506-510

Wood, D.F. and Cornish, K. (2000) Microstructure of Purified Rubber Particles, International Journal of Plant Sciences, Vol. 161, No. 3:435-445

Yew, F.K., Sivakumaran, S. and Chong, D.T. (1990) RRIMFLOW System of Exploitation: Recent Improvements and Update on Yield Performance, RRIM: 42-65

Yoshida, S. (1994) Low Temperature-Induced Cytoplasmic Acidosis in Cultured Mung Bean (*Vigna radiata* [L.] Wilczek), Plant Physiology, Vol. 104, No. 4:1131-1138

THE USE OF SYNERGISTIC HORMONE NUTRIENT COMBINATIONS TO INCREASE THE GROWTH AND QUALITY OF BERRIES

Robert Shortell and Jerry Stoller

Director of Technology Transfer and Chief Executive Officer, respectively, of Stoller Enterprises Inc.; 4001 W. Sam Houston Pkwy. N. STE 100; Houston, TX

Berry crops are significant revenue sources for many agricultural communities. They require a large initial investment and patience to produce a monetary return. Once fruitful, their perennial nature allows them to be harvested for many years, therefore the ultimate return per acre and crop value is a function of yield and fruit quality over time. The objective of this study was to evaluate the effects of synergistic hormone nutrient combinations on growth, longevity and fruit quality parameters on a range of berry crops. Grapes (table and wine) and cherries were evaluated in a randomized complete block design from 2012 through 2014 in Australia, Chile and the United States with and without the application of synergistic combinations of nutrients, kinetin, GA_3 and IBA. Treated and untreated blocks were evaluated for growth, yield and fruit quality parameters both in-season and at harvest. There was a significant increase in tree and vine health, longevity, seasonal growth, yield and fruit quality across crops and locations. These data indicate that hormone nutrient combinations, when applied at the correct timing, can increase return on investment for berry producers.

THE MODE OF ACTION OF AMINOETHOXYVINYLGLYCINE (AVG) FOR FRUIT SET IMPROVEMENT IN TREE FRUIT AND NUT CROPS

Jozsef Racsko, Peter D. Petracek, Rob Fritts Jr., Rick Hopkins

Valent BioSciences Corporation, 870 Technology Way, Libertyville, IL, USA

The plant hormone ethylene is intimately involved in many plant processes including dormancy, germination, flowering, fruit set, general plant growth, maturation, ripening, senescence and abscission. The ability to regulate these ethylene-mediated processes can be a powerful tool for management of crop production and quality. Aminoethoxyvinylglycine (AVG) is a naturally occurring non-protein amino acid that temporarily inhibits ethylene biosynthesis in plants by competitively inhibiting ACC synthase, a key enzyme in ethylene biosynthesis. Since 1997 AVG has been registered primarily as a harvest management tool in pome and stone fruit in several countries of the world. Well established benefits include improved harvest management, reduction of pre-harvest fruit drop, and maintenance of fruit firmness at harvest and through storage. Continuing research has refined application conditions and extended use pattern. Given the importance of ethylene in flowering and fruit set, there has been a considerable interest in the effects of AVG applications on these physiological processes. Depending on the species, there are different modes of action responsible for increased fruit set. For instance, in walnut, nut-producing pistillate flowers abscise as a result of over-pollination and ethylene production, leading to poor yields. Studies have demonstrated that AVG application during flowering improves nut set and yield in walnut. In apples, pears, and cherries AVG increases longevity of floral organs thereby providing a better chance for pollination and fertilization. This presentation will discuss factors (e.g., stigma viability, ovule longevity, pollen tube growth) of AVG effect on fruit set improvement. Since AVG application is done during flowering, presenter will also discuss the impact of AVG on floral attractivity and foraging behavior of insect pollinators.

A NEW APPROACH FOR PRODUCTIVITY OF SOYBEAN AND CORN CROPS

Jerry Stoller, Jun Wei, Albert Liptay, David Lankford

Stoller Enterprises Inc., 4001 W Sam Houston Pkwy N, Houston, TX

Plant growth and development is an array of physiological processes of capturing, converting, consuming, transporting and storing energy. We hypothesize that the temperature difference (ΔT) between plant organs (root, stem, leaves, flowers and fruits) is one of the simple indexes to reflect the energy flux and dynamics (or thermodynamics) among these organs. ΔT could be also related to other physiological processes including MEP pathway that is associated with phytohormones production and activities. Here we present two parts of our research in this area: (1) how we accurately measured the temperature of plant organs; (2) how ΔT could be related to MEP pathway, or hormones production and activities on the circadian basis. Temperature data was collected from corn and soybean plants in open fields every minute and relationship of ΔT and MEP pathway was explored using theoretical and experimental approaches. Our preliminary results showed that ΔT between plant organs can be measured accurately.

9, 10-KETOL-OCTADECADIENOIC ACID (KODA) INFLUENCES FLOWER BUD FORMATION AND MDTFL1 GENE IN APPLES

Satoru Kondo and Hitoshi Ohara

Graduate School of Horticulture, Chiba University, Matsudo 271-8510, Japan

ABSTRACT

Changes of endogenous 9, 10-ketol-octadecadienoic acid (KODA) concentrations, which is synthesized from linolenic acid by 9-lipoxygenase, were analyzed in apple [*Malus sylvestris* (L.) Mill. var. *domestica* (Borkh.) Mansf.] buds. In addition, the effects of KODA application on flower bud formation and the expression of *MdTFL1* (*terminal flower 1*) and *MdFT1* (*flowering locus t 1*) genes in apple buds were investigated in heavy-crop treatment (HCT). An increase of endogenous KODA in the buds in the fruit-thinning treatment (FTT), which resulted in a higher proportion of flower bud formation than in HCT, was observed at 63 days after full bloom, but no such increase was found in HCT. In the heavy-crop trees, the expression of *MdTFL1* in the buds to which KODA was applied was lower than that in untreated buds. These results suggest that endogenous KODA may be associated with flower bud formation, and its application may be effective at improving the proportion of flower bud formation through its effect on *MdTFL1*.

INTRODUCTION

The flower inductions of duckweed (*Lemna paucicostata* L.) and Japanese morning glory (*Pharbitis nil* Choisy) were shown to be associated with KODA levels (Yokoyama et al., 2000). Studies on the molecular mechanisms of flowering have been well established in arabidopsis [*Arabidopsis thaliana* (L.) Heynh. (Bradley et al., 1997)]. Overexpression of *LFY, FT*, or *AP1* has been shown to shorten the vegetative phase, causing early flowering (Kardailsky et al., 1999), whereas overexpression of *TFL1* has been shown to cause late flowering in transgenic arabidopsis (Ratcliffe et al., 1998). It is possible that KODA application induces flower bud formation in apple through an effect on *MdTFL1* transcription. An investigation of the relationship among KODA, *MdTFL1*, and *MdFT1* under various growing conditions may clarify the roles of KODA. However, the effect of KODA application under non-inductive conditions has not been examined. This study examined the effect of KODA application on the transcription of *MdTFL1* and *MdFT1* and flower bud formation in apple under heavy crop load conditions.

RESULTS AND DISCUSSION

The proportion of flower bud formation of flower thinning treatment (FTT) (81.3 ± 3.8%) was significantly higher than that of heavy crop treatment (HCT) (60.3 ± 0.6%). A significant increase in endogenous KODA in FTT was observed at 63 and 103 DAFB, although there was no significant difference of KODA levels at 21 and 49 DAFB between HCT and FTT (Fig. 1). In HCT trees, endogenous KODA levels did not show significant changes from 21 to 103 DAFB. The relative expression levels of *MdTFL1* in $HCT^{+}KODA^{+}$

(HCT with KODA application) were significantly lower than those in HCT$^+$KODA$^-$ (HCT without KODA application) at 21, 35, and 49 DAFB (Fig. 2). In both treatments, the expression of *MdTFL1* decreased sharply after 63 DAFB and leveled off at 77 DAFB in HCT$^+$KODA$^+$ and HCT$^+$KODA$^-$ at 98 DAFB. The expression patterns of *MdFT1* were similar in both KODA treated and -untreated buds under HCT conditions, although the times of the increases and decreases differed (Fig. 2). That is, the expression of *MdFT1* in HCT$^+$KODA$^-$ increased toward 49 DAFB and then decreased at 63 DAFB, whereas that in HCT$^+$KODA$^+$ increased toward 35 DAFB and then decreased at 49 DAFB. The proportion of flower bud formation in HCT$^+$KODA$^+$ (67.6 ± 0.6%) was significantly higher than that in HCT$^+$KODA$^-$ (60.3 ± 0.6%).

The endogenous KODA concentrations in FTT were significantly higher than those of HCT. Furthermore, these concentrations increased significantly at 63 DAFB and before flower bud formation (70 DAFB in this study). This fact suggests that endogenous KODA may play a positive role in flower bud formation in apple. Either endogenous gibberellins (GAs) from the seeds or GA spray is known to inhibit flowering in apple (Ramirez et al., 2001). Therefore, it is possible that low endogenous KODA concentrations in the buds of HCT are associated with high GA concentrations compared with those in FTT (Kittikorn et al., 2010). However, KODA application to the buds did not influence endogenous GA concentrations in the buds (Kittikorn et al., 2011). Therefore, the relationship between auxin and cytokinin, which influence flower bud formation (Sanyal and Bangerth, 1998), and KODA should also be examined. In addition, the origin of endogenous KODA in the buds of apple requires further investigation. A homolog of the *FT* and *TFL1* gene family has been isolated and characterized in fruit trees such as apple (Kotoda and Wada, 2005) and grape (Carmona et al., 2007). *MdTFL1* gene in apple, which is a *TFL1*-like gene, functions equivalently to *TFL1* expressed constitutively in transgenic arabidopsis plants, suggesting that it has the potential to complement the *TFL1* function (Kotoda et al., 2006). It could function redundantly as a flowering repressor and a regulator of vegetative meristem identity because *MdTFL1* is detected strongly in vegetative tissues (Mimida et al., 2009). Our previous report (Kittikorn et al., 2011) showed that the accumulation of *MdTFL1* mRNA transcript levels was found in buds under both heavy-crop and shade treatments that related to the decrease of flower bud formation. The proportion of flower bud formation in apical buds of apple increased after 70 DAFB (Kittikorn et al., 2010). In our study, KODA application in HCT decreased *MdTFL1* mRNA transcript levels at 21, 35, and 49 DAFB, although the expression of *MdTFL1* in HCT$^+$KODA$^-$ decreased sharply after 63 DAFB. It has been shown that *MdTFL1* mRNA was expressed strongly in June, which is 7 weeks after full bloom, which occurs 1 to 2 weeks after the initiation of floral bud formation; from June, the expression decreased gradually toward late July (Kotoda and Wada, 2005). These results support that the induction of flower bud formation and floral development may be coincident with the interruption of *MdTFL1* expression (Kotoda and Wada, 2005). Contrary to the functional role of *TFL1*, *FT* promoted flowering by altering the expression of other endogenous genes (Esumi et al., 2007). In apple, *MdFT1* and *MdFT2* genes have been isolated and their functions clarified (Kotoda and Wada, 2005). A transgenic experiment revealed that both *MdFT1* and *MdFT2* had the potential to function as floral promoters in apple. However, the transcriptional activation of *MdFT1* is more closely associated with the regulation of flower induction in apple than that of *MdFT2*. Therefore, we examined the transcription level of *MdFT1*. In our study, the expression of *MdFT1* in HCT+KODA+ reached a peak earlier and higher than in HCT+KODA-. It has been reported that the increase of *MdFT1* (Kotoda et al., 2010) and *AFL1* and *AFL2* transcription levels (apple orthologs of *LFY*) (Wada et al., 2002) in the apical meristems of one-year-old apple shoots began 1 to 2 weeks earlier than flower initiation. The increase of the proportion of flower bud formation at 70 DAFB (Kittikorn et al., 2010), which did not occur after the increase of *MdFT1* but did occur after the decrease of *MdTFL1* expression, suggests that the role of *MdTFL1* may be downstream of *MdFT1*. It was demonstrated that the dynamic state of *MdFT1* and *MdTFL1* genes observed in our study may be associated with flower bud formation.

LITERATURE CITED

Bradley, D., O. Ratcliffe, C. Vincent, R. Carpenter, and E. Coen. 1997. Inflorescence commitment and architecture in Arabidopsis. Science 275:80–83.

Carmona, M.J., M. Calonje, and J.M. Martı´nez-Zapater. 2007. The *FT/TFL1* gene family in grapevine. Plant Mol. Biol. 63:637–650.

Esumi, T., R. Tao, and K. Yonemori. 2007. Relationship between floral development and transcription levels of the leafy and termanal flower 1 homologs in japanese pear (*Pyrus pyrifolia* Nakai) and quince (*Cydonia oblonga* Mill.). J. Jpn. Soc. Hort. Sci. 76:294–304.

Kardailsky, I., V.K. Shukla, J.H. Ahn, N. Dagenais, S.K. Christensen, J.T. Nguyen, J. Chory,M.J. Harrison, and D. Weigel. 1999. Activation tagging of the floral inducer FT. Science 286:1962–1965.

Kittikorn, M., K. Okawa, H. Ohara, N. Kotoda, M. Wada,M. Yokoyama, O. Ifuku, and S. Kondo. 2011. Effect of fruit load, shading, and 9, 10-ketol-octadecadienoic acid (KODA) application on *MdTFL1* and *MdFT1* genes in apple buds. Plant Growth Regulat. 64:75–81.

Kittikorn, M., N. Shiraishi, K. Okawa, H. Ohara, M. Yokoyama, O. Ifuku, S. Yoshida, and S. Kondo. 2010. Effect of fruit load on 9, 10-ketol-octadecadienoic acid (KODA), GA and jasmonic acid concentration in apple buds. Sci. Hort. 124:225–230.

Kotoda, N., H. Hayashi, M. Suzuki,M. Igarashi, Y. Hatsuyama, S. Kidou, T. Igasaki, M. Nishiguchi, K. Yano, T. Shimizu, S. Takahashi, H. Iwanami, S. Moriya, and K. Abe. 2010. Molecular characterization of flowering locus t-like genes of apple (*Malus×domestica* Borkh.). Plant Cell Physiol. 51:561–575.

Kotoda, N., H. Iwanami, S. Takahashi, and K. Abe. 2006. Antisense expression of *MdTFL1*, a *TFL1*-like gene, reduces the juvenile phase in apple. J. Amer. Soc. Hort. Sci. 131:74–81.

Kotoda, N. and M. Wada. 2005. *MdTFL1*, a *TFL1*-like gene of apple, retards the transition from the vegetative to reproductive phase in transgenic arabidopsis. Plant Sci. 168:95–104.

Mimida, N.,N.Kotoda, T. Ueda,M. Igarashi,Y. Hatsuyama, H. Iwanami, S. Moriya, and K. Abe. 2009. Four TFL1/CEN-like genes on distinct linkage groups show different expression patterns to regulate vegetative and reproductive development in apple (*Malus ×domestica* Borkh.). Plant Cell Physiol. 50:394–412.

Ramirez, H., G.V. Hoad, A. Benavides, and E. Rangel. 2001. Gibberellins in apple seeds and the transport of [^{3}H]-GA_4. J. Mexican Chem. Soc. 45:47–50.

Ratcliffe, O.J., I. Amaya, C.A. Vincent, S. Rothstein, R. Carpenter, E.S. Coen, and D.J. Bradley. 1998. A common mechanism controls the life cycle and architecture of plants. Development 125:1609–1615.

Sanyal, D. and F. Bangerth. 1998. Stress induced ethylene evolution and its possible relationship to auxin-transport, cytokinin levels, and flower bud induction in shoots of apple seedlings and bearing apple trees. Plant Growth Regulat. 24:127–134.

Yokoyama, M., S. Yamaguchi, S. Inomata, K. Komatsu, S. Yoshida, T. Iida, Y. Yokokawa, M. Yamaguchi, S. Kaihara, and A. Takimoto. 2000. Stress-induced factor involved in flower formation of lemna is an a-ketol derivative of linolenic acid. Plant Cell Physiol. 41:110–113.

Wada, M., Q. Cao, N. Kotoda, J. Soejima, and T. Masuda. 2002. Apple has two orthologues of floricaula/leafy involved in flowering. Plant Mol. Biol. 49:567–577.

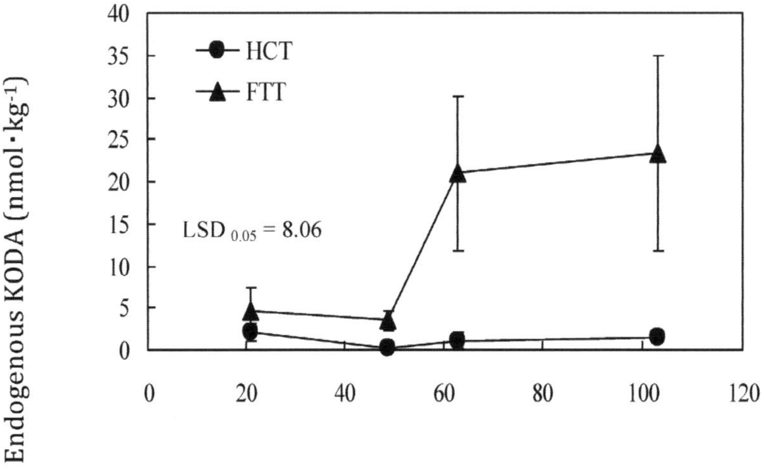

Fig. 1. Changes in endogenous 9, 10-ketol-octadecadienoic acid (KODA) concentrations in the apical buds of 'Fuji' apple subjected to the flower thinning treatment (FTT), in which all flowers were removed at full bloom and heavy crop treatment (HCT)

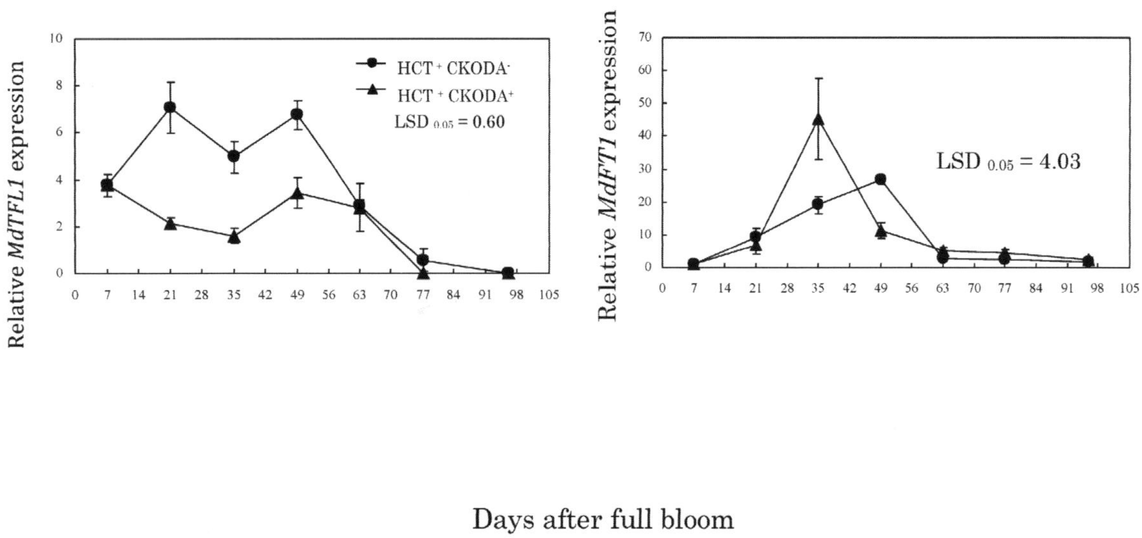

Fig. 2. Quantitative real-time reverse transcription–polymerase chain reaction (RT-PCR) analysis of *MdTFL1* and *MdFT1* in apical buds of 'Fuji' apple treated with KODA under heavy-crop treatment (HCT).

IMPROVING GRAIN YIELD IN RICE USING PLANT HORMONE INHIBITORS

Hiroaki Tamaki[1,2], Maria Reguera[1], Yasser M. Abdel-Tawab[1], Yumiko Takebayashi[3], Hiroyuki Kasahara[3], Eduardo Blumwald[1]

[1]*Department of Plant Sciences, University of California, Davis, California 95616, United States of America*

[2]*Health and Crop Sciences Research Laboratory, Sumitomo Chemical Co. Ltd., Hyogo 665-8555, Japan*

[3]*RIKEN Center for Sustainable Resource Science, Yokohama, Kanagawa 230-0045, Japan*

Developing high grain-yielding rice varieties is of paramount importance to fulfill the world's demand of food. Among different rice varieties, several genes associated with grain yield improvement have been identified by QTL analysis. Most of the genes identified have been functionally associated with sink strengthening, and the mechanisms mediating source strengthening that result in yield improvement are still largely unknown. Using a chemical approach, we aimed at the characterization of hormone-related mechanisms responsible for yield enhancement through the regulation of source function in rice.

Among the tested plant hormone inhibitors (Fig. 1), nordihydroguaiaretic acid (NDGA; abscisic acid (ABA) biosynthesis inhibitor) and 2-aminoisobutyric acid (AIB; ethylene biosynthesis inhibitor) improved grain yield in a concentration dependent manner (Figs. 2A and 2B).

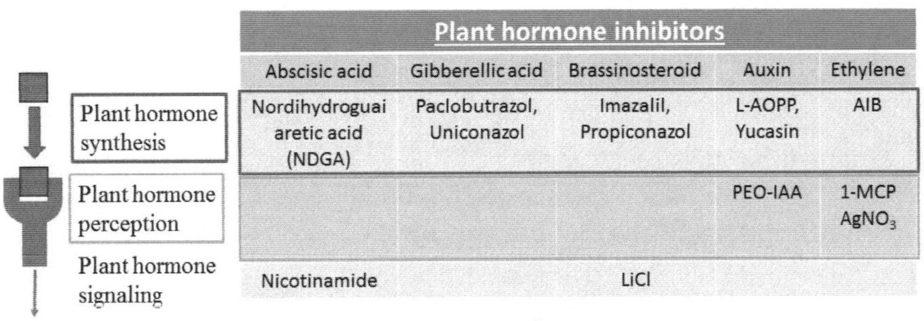

Fig 1. Plant hormone inhibitors for biosynthesis, perception, and signaling

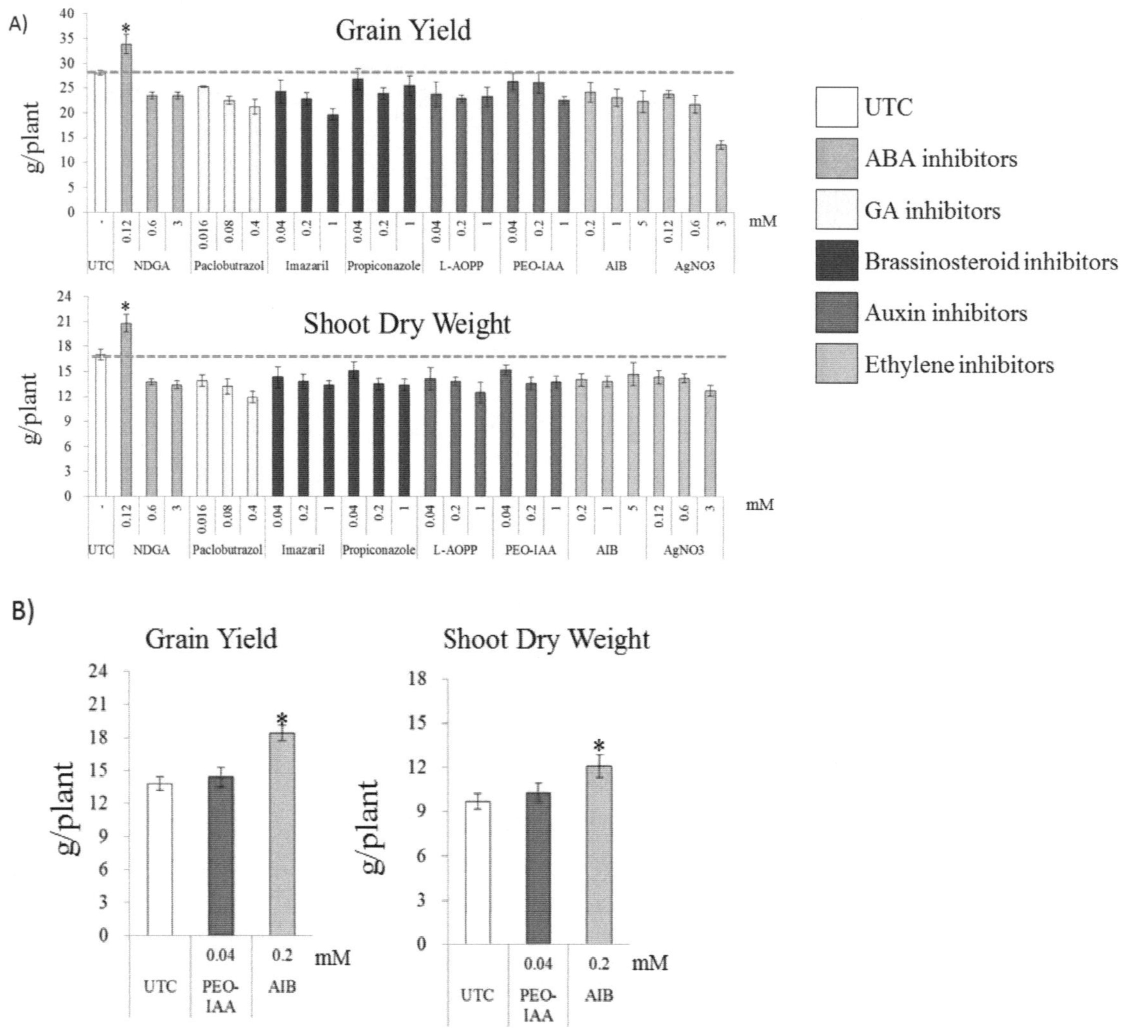

Fig 2. Effects of plant hormone inhibitors on biomass

A) Effects of plant hormone inhibitors sprayed at pre-anthesis
B) Effects of plant hormone inhibitors sprayed at post-anthesis

NDGA treatment increased photosynthesis and the starch content in flag leaves at pre-anthesis, and increased starch contents in grains after harvest (Fig. 3). At the concentration indicating the effect on grain yield improvement, NDGA inhibited the expression of ABA-responsive genes, but did not significantly decreased ABA content. However, NDGA significantly decreased the contents of jasmonic acid and jasmonic acid-isoleucine. Co-application of NDGA and Methyl-Jasmonate did not show any improvement in photosynthesis and grain yield (Fig. 4). These results strongly suggested that NDGA increased source function by decreasing the contents of jasmonic acid and jasmonic acid-isoleucine to increase grain yield in rice. Taken together, our results support the notion that the specific inhibition of jasmonic acid or ethylene biosynthesis resulted in grain yield improvements in rice.

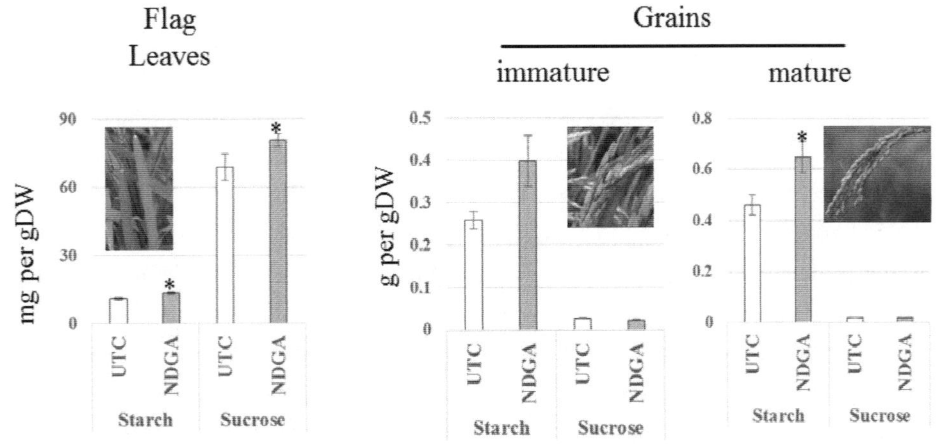

Fig 3. Contents of starch and sucrose in flag leaves and grains

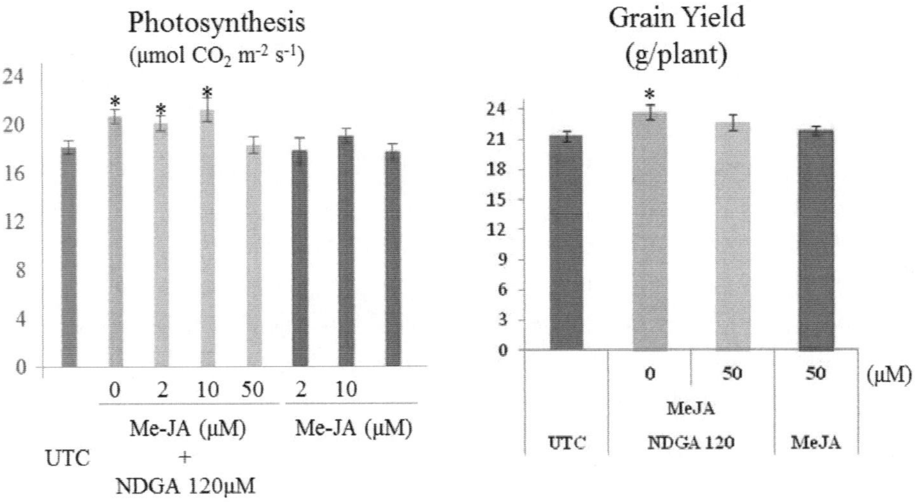

Fig 4. Photosynthesis and grain yield in the treatment of Me-JA and/or NDGA

UNDERSTANDING NUT GROWTH PHENOLOGY AND CHILL ACCUMULATION IN PISTACIA VERA

Cara Allan, University of California, Davis

Louise Ferguson and Gurreet Brar, University of California Cooperative Extension

Studies were conducted to document pistachio nut growth stages as a function of thermal units. Hourly temperatures at six locations throughout California's San Joaquin Valley were recorded to track thermal unit (TU) accumulations and at these locations 2-5 different cultivars (Kerman, Kalehghouchi, Pete I, Lost Hills and Golden Hills) were sampled on a weekly basis starting with fruit set. Nut growth from that point on was plotted as a function of the accumulation of TUs with two different models and the parameters for these three different fruit stages were determined: i) growth of pericarp to its final size, ii) hardening of the endocarp, and iii) growth of embryo. The number of thermal units to reach and complete each of these stages was calculated. In the non-linear model Stage I requires 670.6 TU to reach the maximum volume of 3544.66 ± 432.34 mm^3, while in DJPheno modeling software the maximum volume of 3350.66 mm^2 was completed at 563 TU. Stage II reaches its completion at 2423.58 TU when the shell withstands a maximum of 23.60 lbs/3mm^2 according to the non-linear model and a maximum of 21.55 lbs/3mm^2 after 1863 TU in DJPheno. Stage III of the non-linear model begins at 913.74 TU when the embryo begins to grow and ends with the embryo reaching a length of 21.53±1.44 mm at 1879.66 TU. In the DJPheno model Stage III begins around 790 TU and ends at 1739 TU. Experiments were also started to study the effects of California's winter weather patterns on pistachio bloom, set and yield. Over the last 33 years the number of winter fog events has decreased by 46% on average (Baldocchi, 2014). In preliminary studies it was found that the bud temperature on the south side of the tree was significantly higher than north side on sunny days, possibly affecting chill accumulation, while there was no significant difference on the foggy days.

EFFECTS OF PROTEIN PHOSPHORYLATION IN HORMONE SIGNALING

Hong-Wei Xu

National Key Laboratory of Plant Molecular Genetics

The casein kinase 1 (CK1) family participates in various cellular processes in eukaryotes, however, whose function in higher plants remains largely unknown. By systemic genetic studies, we demonstrated that CKI members regulate gibberellin (GA) signaling and ethylene biosynthesis through phosphorylating and regulating the stability of target proteins, providing insights into the regulatory mechanism of hormone signaling.

POLYMERIZATION AS A HIDING STRATEGY FOR THE MANAGEMENT OF PLANT SOURCE-SINK RELATIONSHIPS

Richard Woodward and Jerry Stoller

Stoller Enterprises, Inc. Houston, TX 77043

Polymerization of carbon, nitrogen and phosphorus converts soluble, metabolically active compounds such as sugars, amino acids and phosphate into polymers such as starch, glutens and phytate. Polymerization of pathway products into feedback-inert compounds effectively limits their potential for feedback inhibition of anabolic processes. Plant growth regulators define sink strength by modulating the rate and extent of polymerization. Exogenous application of cytokinin, and to a lesser extent gibberellic acid, enhances sink strength, auxin is neutral, and ethylene reduces sink strength.

The cytoplasmic concentration of phosphorus is a surrogate for energy flux in plants. Energy flows towards phosphorus deficient tissues. Phosphorus gradients are defined by uptake, transport, sequestration and polymerization. To attract energy in the form of carbohydrates, storage tissues maintain the appearance of phosphorus deficiency by removing soluble phosphate through polymerization as phytate; a saturated hexaphosphate carbohydrate ring. Enhancing phytate synthesis enhances sink strength and subsequent yield.

Phytate is the major source of phosphate in the manure of non-ruminants (fowl, swine, equine, etc.). Manure application increases the concentration of phytate in soils. Phytate is recalcitrant to soil microbial degradation, provides limited plant-available phosphorus and accumulates in manure-amended fields. Without supplemental inorganic phosphorous, yields decline even though soils test high for phosphorus. Environmental concerns increasingly limit the application of phosphate fertilizers. The addition of selected micronutrients [essential for soil microbes] in starter fertilizers stimulate and support the rapid growth of phytate degrading microbes. This liberates bound phosphate for root uptake and helps reduce the accumulated phosphate in manured soils.

Stoller technology provides opportunities for the management of both the synthesis and degradation of phytate for yield enhancement and phosphorus use efficiency respectively.

A SMALL MOLECULAR PEPTIDE DERIVED FROM ANIMAL SERUM ALBUMIN REGULATE PLANT GROWTH

Hao Li[1,2]

1 College of Biological Science and Engineering, Fuzhou University, China, 350108

2 Department of Medicine, The University of Chicago, Chicago, IL, 60637

Not only endogenous peptide from plant can regulate plant growth. Exogenous peptide with small molecule derived from animal serum albumin promote plant root germination and growth. Here we report that a peptides with seven amino acids, named SI-7, were determined the effects on promoting significantly growth using different kinds of plants of fruit, corn, vegetable and flower. Peptide SI-7 with lower concentration of 10 nM promoted the seeds of watermelon, rice, flower, spinach growth and the simplified vigor increased rate compared with untreated control were 82.40%, 52.03%, 57.92% and 30.60%, respectively; the increase rates of root length were 57.47%, 69.07%, 7.91% and 46.48%; the increase rates of seeding length were 135.94%, 91.08%, 5.40% and 31.69%. The mechanism of promoting plant growth were explored preliminarily. The ultrastructure of seeds treated by SI was investigated by transmission electron microscopy method. It showed that radicle cell contents and various organelles of watermelon seed were clearly visible to have many rough endoplasmic reticulums associated with protein synthesis, Golgi apparatus related to protein's processing and packaging and polysaccharide synthesis, and mitochondrion involved in providing energy for the organism in radicle cells. The change of protein content, SOD, POD, CAT activity and endogenous hormone IAA, GA_3 and ABA content of watermelon seeds treated by SI-7 during germination were tested, respectively. The highest protein content was up to 0.587 g/L. The highest SOD activity was 228.92 U/mg. The highest POD activity was 27.04 U/mL. The highest CAT activity were 21.01 U/mg. The results showed that the small molecular peptide promoted significantly plant growth with lower concentration.

EFFECTS OF NAA AND BA ON GRAFT HEALING AND PLANT GROWTH RESUMPTION OF GRAFTED TOMATO SEEDLINGS

Bizhen Hu, Joshua Blakeslee, Matt Kleinhenz, Department of Horticulture and Crop Science, The Ohio State University, Ohio Agricultural Research and Development Center, Wooster, OH, USA

Peter Ling, Department of Food, Agricultural and Biological Engineering, The Ohio State University, Ohio Agricultural Research and Development Center, Wooster, OH, USA

Mark Bennett, Department of Horticulture and Crop Science, The Ohio State University, Columbus, OH, USA

INTRODUCTION

Grafting can improve vegetable production by quickly combining the desirable traits of two varieties into a single plant representing a physical hybrid. However, the development of improved, grafted planting stock is possible only after newly-grafted plants heal wounds created during the grafting process. Likewise, once healed, grafted plants must resume growth and withstand mechanical and other stresses during and after planting. The efficiency of grafting operations is often assessed based on the percentage of successful grafts, the quality (e.g., mechanical strength) of rootstock-scion connections and the time and other resources required to develop planting-ready stock. Inexpensive and reliable steps to increase the efficiency of grafted plant preparation will facilitate wider use of the process and grafted planting stock.

Histological evidence indicates that healing of newly grafted seedlings includes two stages: cell division, resulting in callus formation, and wound vessel differentiation, resulting in rootstock-scion vascular reconnection (Jeffree and Yeoman, 1983; Lindsay et al., 1974; Martínez-Ballesta et al., 2010). Vascular reconnection linking rootstock and scion is generally regarded as the completion of healing. However, further change may be required, particularly if newly reconnected root-shoot axes are to be considered as functional and resilient as root-shoot axes in healthy, intact, ungrafted seedlings. Direct, visual, tissue-level evidence suggests that vascular reconnection is often achieved approximately seven days after grafting (Fernández-García et al., 2004).

Phytohormones, especially auxin and cytokinin, mediate healing (Köse and Güleryüz, 2006; Yin et al., 2012). For example, auxin accumulated at the graft interface as cell division began and elevated levels were maintained through vessel differentiation until vasculature reconnection (Yin et al., 2012). Additional descriptions of the role of phytohormones in grafted plant preparation would be beneficial.

We set out to document the effects of exogenous applications of auxin and cytokinin on the healing and regrowth of newly grafted tomato seedlings. Reports of previous similar tests appear to be unavailable. Our two studies involved five concentrations of 1-Naphthaleneacetic acid (NAA) or 6-benzylaminopurine (BA) applied to the cut surface of tomato rootstock hypocotyl sections during self-grafting, followed by measures of vasculature reconnection and shoot development. Erythrosine B (a translocatable dye) was used to visualize rootstock to scion water flow and eleven measures of the reconnected root-shoot axis were taken to track plant growth.

METHODS

Two studies were completed at The OSU-OARDC in Wooster, OH from April to May, 2015. Study 1 included nine treatments: 0.1, 1, 10 mg/L NAA; 0.1, 1, 10 mg/L BA; 0.1 mg/L NAA + 1 mg/L BA, 1 mg/L NAA + 1 mg/L BA, and 1 mg/L NAA + 10 mg/L BA. Study 2 had six treatments: 10, 100, 1000 mg/L NAA; 10, 100, 1000 mg/L BA. A water-only control was employed in both studies. NAA and BA (Sigma-Aldrich, St Louis, MO) were dissolved in a few drops of 1 M potassium hydroxide, and 1 M hydrochloric acid, respectively, and then doubled distilled water was added to make a 1 mg/ml stock solution of each compound. The stock solution was stored at 4 °C before being diluted to target concentrations at the time of use.

Plant materials used in both studies were 'Cherokee Purple' tomato. Seedlings were grown in an environmental controlled greenhouse with temperature set as 22-26 °C from dawn to dusk, and 17-21 °C from dusk to dawn. High-intensity discharge lights came on at dawn and stayed on for 16 h unless outdoor sunlight stayed above 350 $w/m^2/s$ for 20 minutes. Seed were sown in 96-cell trays preloaded with Pro-Mix® MP Mycorrhizae™ Organik™ (Premier Tech, Rivière-du-Loup, Québec, Canada) growing medium. Trays were placed on Kapmat (Buffalo Felt Products Corp, NY) underlain by 4-millimeter thick plastic on a 5.4 m x 1.8 m bench in the greenhouse. Trays were hand-misted to wetness immediately after sowing; a bench-top, automated irrigation system was used thereafter for 1 week to maintain soil moisture, for each tray, with 1 fogger pulsed on for 8 s every 30 min every day and 2 drippers on the Kapmat pulsed on for 3 min at 7 am, 11 am and 3 pm every day. One week after seed sowing, seedlings were moved to another bench without automatic irrigation systems and manually watered once a day.

Tomato seedlings were splice grafted three weeks after sowing, with 1 µl of treatment solution applied to the cut surface of rootstock hypocotyl sections using a 5 µl syringe. Graft unions were secured by commercial clips and plants were transferred to a healing chamber consisting of a polyvinyl chloride (PVC) frame covered by one layer of clear plastic sheeting and two layers of black knitted shade cloth (50% PAR transmission) and located in a climate-controlled greenhouse. The temperature was 26±5 °C and relative humidity was 70±9% in the healing chamber during the study period. (Fig. 1)

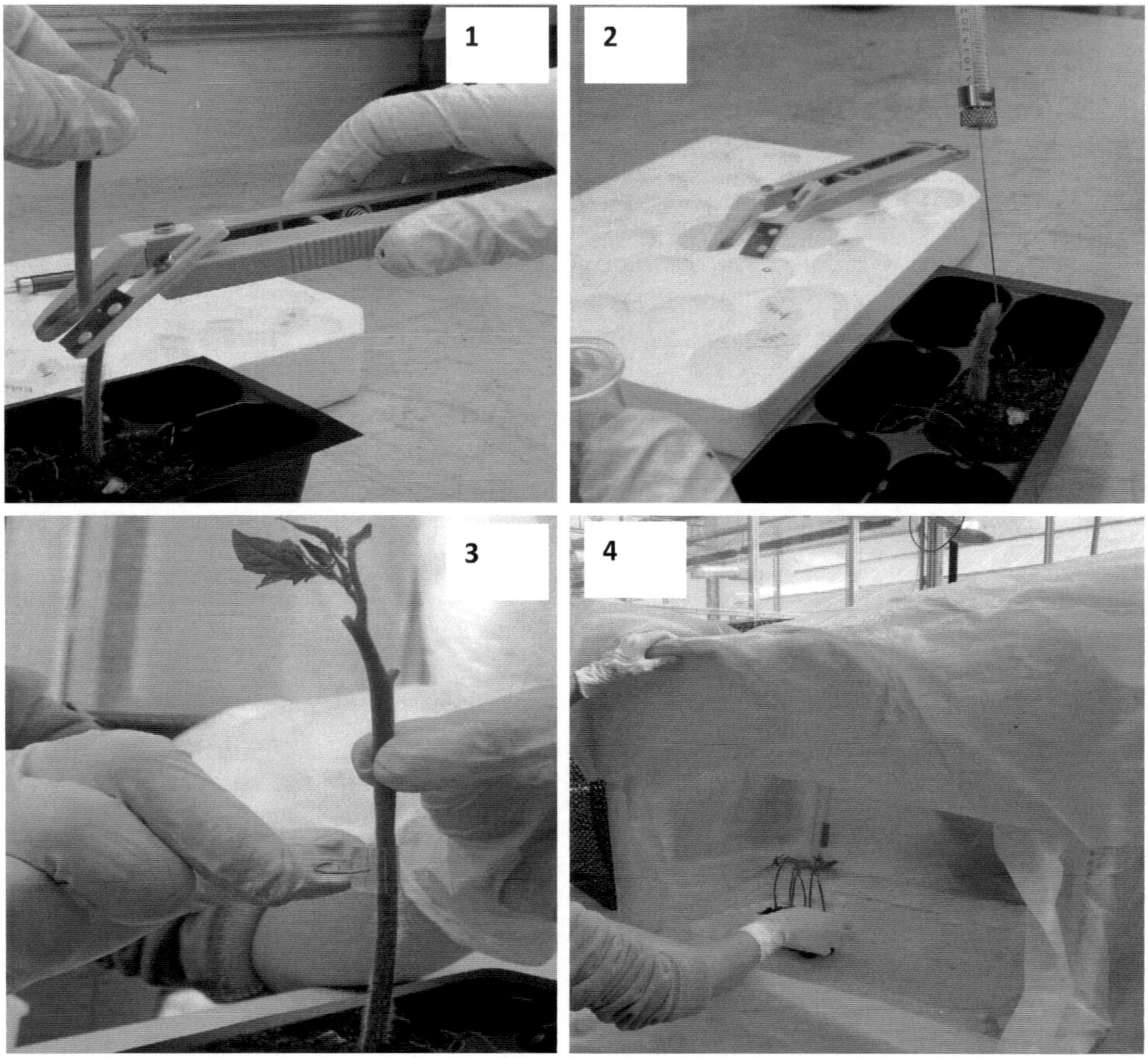

Fig. 1. Grafting and treatment application. Beginning the grafting process (1); Application of 1 µl NAA or BA solution to the cut surface of rootstock hypocotyl sections (2); Securing the graft union with a commercial clip (3); Placing grafted plants in a healing chamber (4).

Graft healing (vasculature reconnection) was assessed based on the movement of Erythrosine B dye across the graft union. Five days after grafting in Study 1 and 4 days after grafting in Study 2, grafted plants were excised at the soil line and placed upright in a 1 mg/ml Erythrosine B solution for 15 min, submerging the cut surface 2 mm. Plants were taken from the solution at 15 min and the epidermis was removed by peeling, allowing for the visualization and measurement of upward dye movement. Distance of dye movement was measured from the soil line and graft union to the endpoint of the dye front in the scion (Fig. 2). Plant growth was monitored 10 days after grafting in both studies by recording the following variables: plant height from the soil line to the meristem, scion length from the graft union to the meristem, stem diameter at the rootstock and scion, leaf area based on digital images analyzed by WinCAM (Regent Instruments, Quebec, QC, Canada) software (Fig. 3), relative growth of the above variables calculated as (final value 10 days after grafting-initial value at grafting)/initial value*100%, fresh and dry weight of leaf and stem, specific leaf area calculated as leaf area/leaf dry weight, and compactness calculated as aboveground dry weight/plant height.

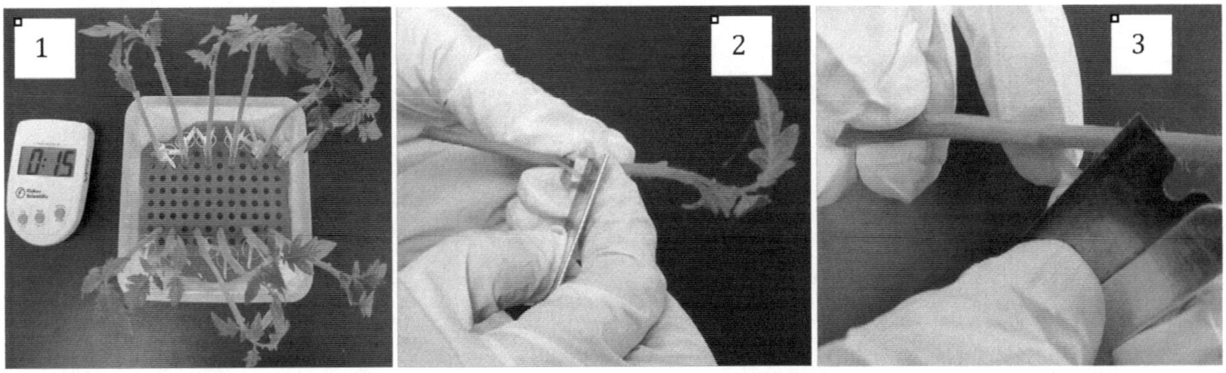

Fig. 2. Dye movement measurement to monitor vasculature reconnection in grafted tomato plants. Plants excised at the soil line were placed in a 1 mg/ml Erythrosine B solution for 15 min (1); Scion stem sections were peeled to observe and measure dye movement (2, 3).

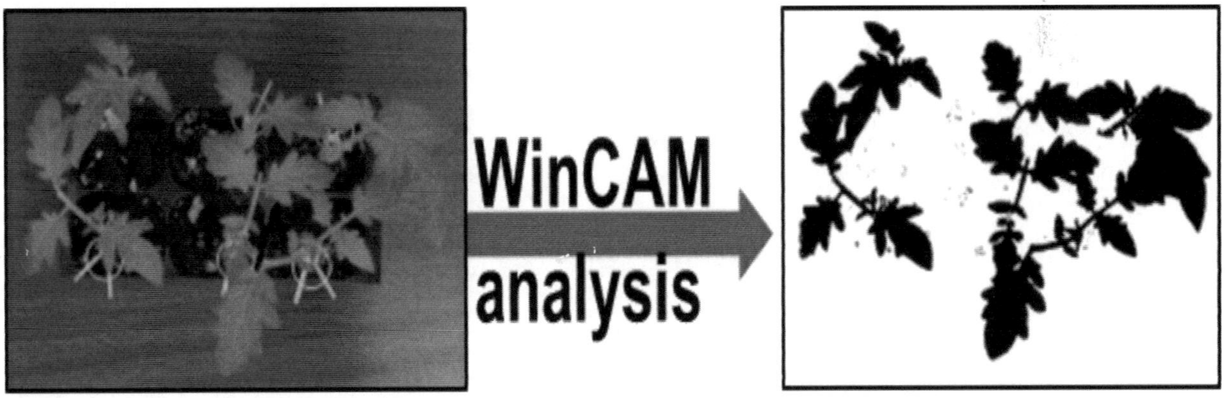

Fig. 3. Digital images of grafted plants taken above the canopy followed by analysis by WinCam software, providing estimates of leaf area. WinCam separates the color(s) of interest (e.g., green) from background and calculates the percentage of the whole area each color represents, providing estimates of leaf area in this application.

Both studies were completed using a randomized complete block design. Study 1 included four hundred grafted plants, two plants per block with ten blocks for vasculature reconnection measurement 5 days after grafting, and two plants per block with ten blocks for plant growth measurement 10 days after grafting. Study 2 included four hundred and twenty grafted plants, six plants per block with five blocks for vasculature reconnection measurement 4 days after grafting, and six plants per block with five blocks for plant growth measurement 10 days after grafting.

Statistical analyses were completed in SAS (version 9.3; SAS Institute, Cary, NC). Treatment effects were analyzed through ANOVA. Multiple comparisons were conducted using LSD tests.

RESULTS

The distance of dye movement did not differ among treatments in either study (Table 1). Of eleven growth variables measured, only leaf area relative growth differed in Study 1, showing larger values in control plants and plants given 0.1 mg/L NAA at grafting (Table 1, Fig. 4). In Study 2, plant height and scion length relative growth decreased with increases in NAA or BA concentration, although rootstock stem diameter relative growth was larger at 1000 mg/L NAA and scion stem diameter relative growth was larger at 1000 mg/L BA (Table 1, Fig. 5). NAA at 10, 100, 1000 mg/L caused undesirable rooting from the graft union (Fig. 6).

Table 1. ANOVA output describing the influence of exogenous NAA and BA on the healing and regrowth of newly grafted tomato seedlings.

Study	Stained length (cm)	Stained length: plant height (%)	Stained length: scion length (%)	Plant height relative growth (%)	Scion length relative growth (%)	RS diameter relative growth (%)	Scion diameter relative growth (%)	Leaf fresh weig ht (g)	Stem fresh weight (g)	Leaf dry weight (g)	Stem dry weight (g)	Comp act- ness (mg/c m)	Leaf area relative growth (%)	Specific leaf area (cm^2/g)
1	NS	NS	NS	NS	NS	NS	NS	NS	NS	NS	NS	NS	0.0135	NS
2	NS	NS	NS	0.0002	0.0003	0.0019	0.0272	NS	NS	NS	NS	NS	NS	NS

NS: no significant difference.

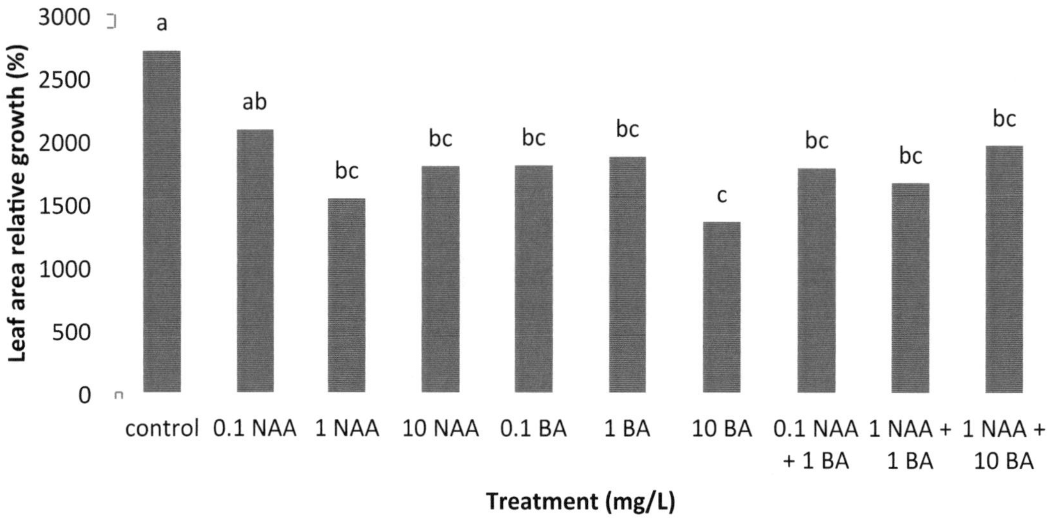

Fig. 4. Leaf area relative growth 10 days after grafting in Study 1. Different letters on bars represent significance at $p<0.05$.

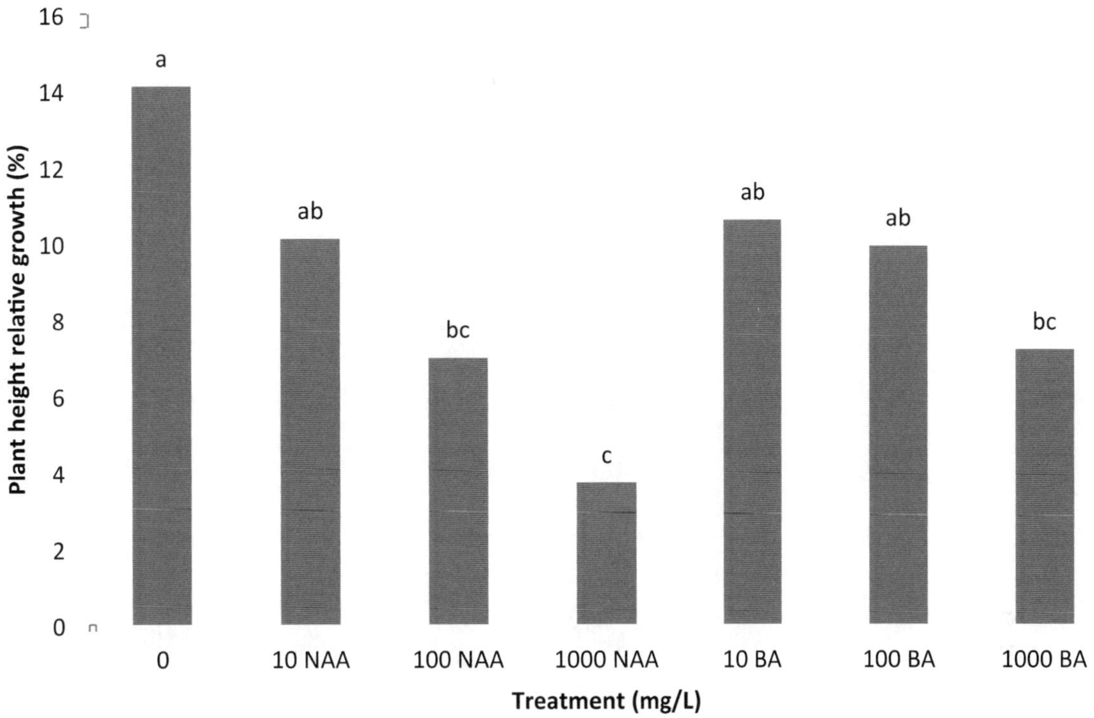
a
ab
bc
c
ab
ab
bc
Plant height relative growth (%)
16
14
12
10
8
6
4
2
0
0
10 NAA
100 NAA
1000 NAA
10 BA
100 BA
1000 BA
Treatment (mg/L)

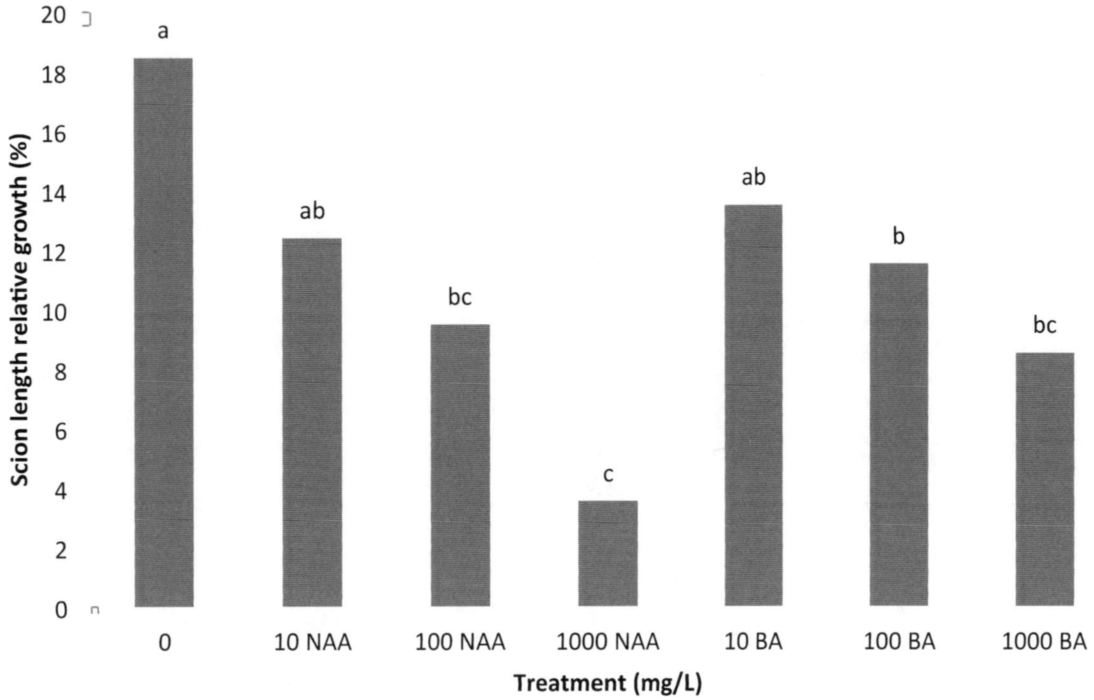
a
ab
bc
c
ab
b
bc
Scion length relative growth (%)
20
18
16
14
12
10
8
6
4
2
0
0
10 NAA
100 NAA
1000 NAA
10 BA
100 BA
1000 BA
Treatment (mg/L)

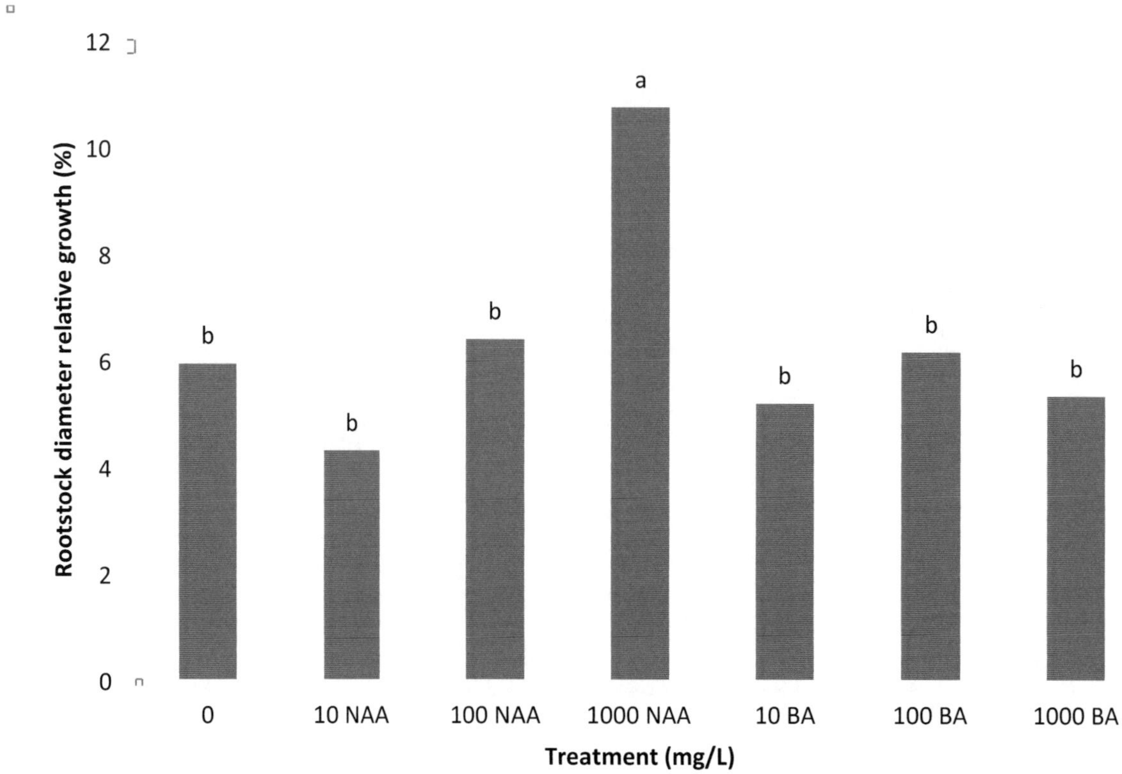

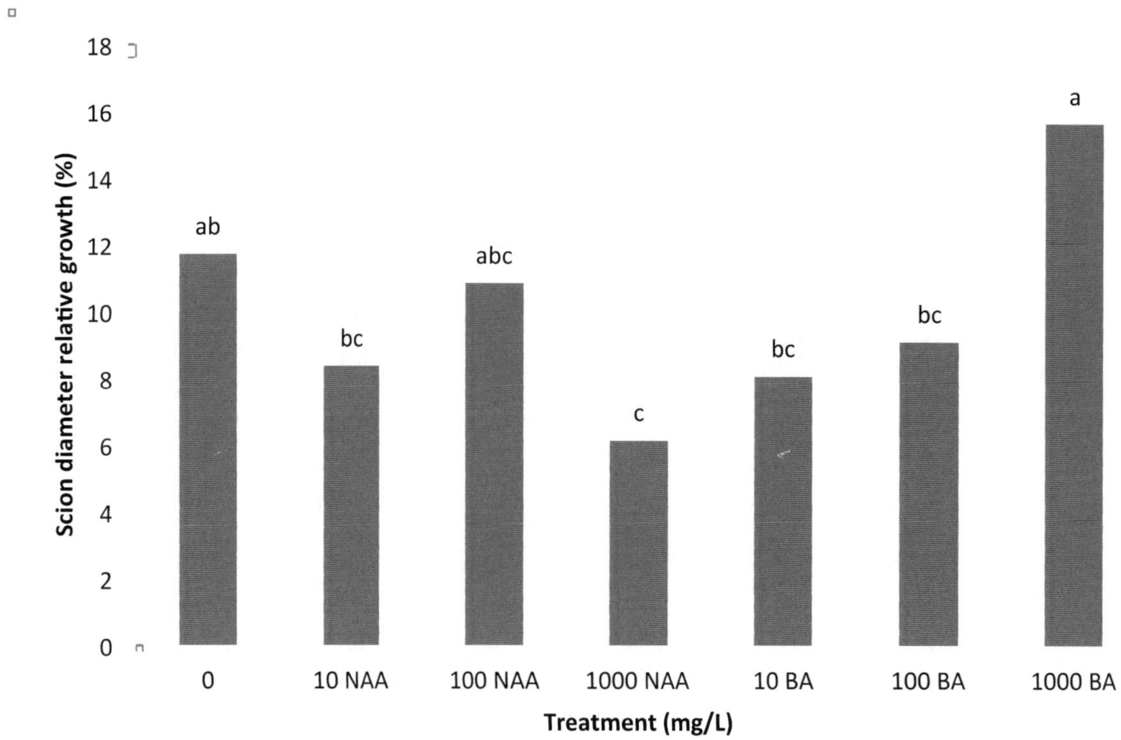

Fig. 5. Relative growth of plant height, scion length, rootstock and scion diameter 10 days after grafting in Study 2. Different letters on bars represent significance at $p<0.05$.

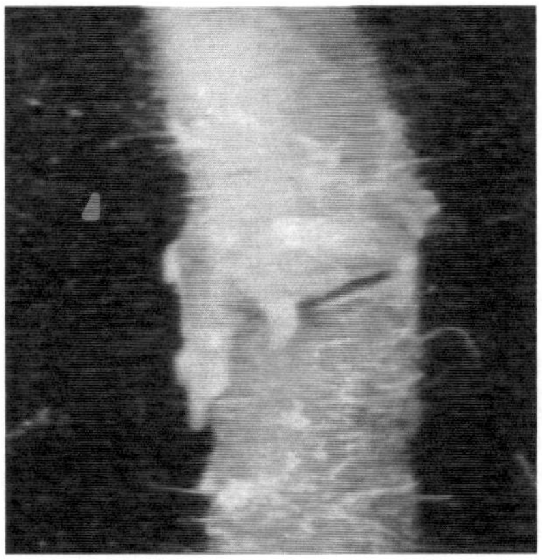

Fig. 6. Undesirable rooting from the graft union following the application of 10, 100, or 1000 mg/L NAA at grafting.

CONCLUSION

Collectively, these results suggest that NAA and BA as used in this study are unlikely to significantly alter healing efficiency (rate of vasculature reconnection, shoot growth) in tomato. Still, other plant growth regulators application parameters warrant further exploration.

ACKNOWLEDGEMENT

Salaries and research support provided in part by State and Federal funds appropriated to the Ohio Agricultural Research and Development Center, The Ohio State University. Support also provided by the USDA-NIFA 2011 Specialty Crop Research Initiative (Award #2011-51181-30963).

LITERATURE

Fernández-García, N., M. Carvajal, and E. Olmos. 2004. Graft union formation in tomato plants: peroxidase and catalase involvement. Ann. Bot. 93(1):53-60.

Jeffree, C. and M. Yeoman. 1983. Development of intercellular connections between opposing cells in a graft union. New Phytol. 93(4):491-509.

Köse, C. and M. Güleryüz. 2006. Effects of auxins and cytokinins on graft union of grapevine (*Vitis vinifera*). New Zealand. J. Crop Hortic. Sci. 34:145–150.

Lindsay, D.W., M.M. Yeoman, and R. Brown. 1974. An analysis of the development of the graft union in *Lycopersicon esculentum*. Ann. Bot. 38:639-646.

Martínez-Ballesta, M.C., C. Alcaraz-López, B. Muries, C. Mota-Cadenas, and M. Carvajal. 2010. Physiological aspects of rootstock–scion interactions. Sci. Hort. 127:112–118.

Yin, H., B. Yan, J. Sun, P. Jia, Z. Zhang, X. Yan, J. Chai, Z. Ren, G. Zheng, and H. Liu. 2012. Graft-union development: a delicate process that involves cell–cell communication between scion and stock for local auxin accumulation. J. Expt. Bot. 63(11):4219-4232.

THE ROLE OF MICROBIOME ON PLANT PHENOTYPES AND CROP YIELDS

Marcus Meadows-Smith

BioConsortia, Inc., Davis, CA 95618 USA

The microbiome has held the spotlight in human health and physiology research for many years, with recent studies finding strong associations between gut bacteria and health, nutrition, immune system, and neurological function. Similarly, in plants, microbes have a profound effect on survival, health, growth and development. Within the soil, on the roots, and in the plant itself, natural microorganisms are providing access to nutrients, protecting against pests and disease, affecting chemical signaling, modifying structure, as well as acting as pathogens.

Some of these microbes have been identified and put to use in agriculture, such as nitrogen-fixing bacterium *Azospirillum* sp., plant growth promoting fungi *Trichoderma* spp., and anti-fungal anti-bacterial *Bacillus subtilis*. In the midst of a growing desire and need for sustainable agricultural growth and heightened regulatory requirements blocking certain synthetics and chemistries, microbes offer a much needed natural solution. This opportunity, however, is hindered by challenges of efficacy and consistency. Advancements in technologies for genome sequencing, big data management, microbiome research, and other 'omics' analyses, have started to improve the scientific foundation behind microbial R&D.

BioConsortia's Advanced Microbial Selection (AMS) technology incorporates insight from plant physiology, microbiome analysis, and computational biology to discover microbial consortia effective at improving targeted plant traits and final yields. The AMS process reverses conventional methodology, putting the plant at the center of the selection process. Using phenotypic indicators of yield at maturity and information on microbial community structure, the AMS process moves plants through iterative rounds of plant-microbe selection, rapidly evolving the microbiome and enriching the community with those microbes enhancing stress resisting and growth improving characteristics within the plant. The resulting teams of complementary microbes provide a range of functional effects, have been identified as good plant colonizers, and, due to inclusion of typical chemistries, work within standard agronomic practice. BioConsortia's growth chamber and lab results offer preliminary evidence of success, showing 10-30% increases over control for a variety of measured parameters. Field trial results in the U.S., E.U. and New Zealand also proved positive, most notably through high levels of consistency within treatments.

Future technological advances and further investment in research will increase our understanding of microbes and bring forth superior biological products. Improvements in consistency and dramatic effects on yield improvement will motivate continued adoption by mainstream agriculture, sustainably improving global food production in the growing world.